HISTORY OF TECHNOLOGY SERIES 2

SERIES EDITOR: B. BOWERS

Early radio wave detectors

Early radio wave detectors

Vivian J. Phillips B.Sc.(Eng.), Ph.D., ACGI, DIC, CEng., FIEE, FIERE
Senior Lecturer in Electrical Engineering, University College of Swansea,
Wales

PETER PEREGRINUS LTD
in association with the
SCIENCE MUSEUM, LONDON

Published by: The Institution of Electrical Engineers, London and
               New York
               Peter Peregrinus Ltd., Stevenage, U.K, and New York

© 1980: Institution of Electrical Engineers

British Library Cataloguing in Publication Data

Phillips, V.J.
    Early radio wave detectors. — (History of technology; 2)
    1. Radio detectors — History
    I. Title   II. Series
    621.3841 '7   TK6565.D4

    ISBN 0-906048-24-9

Printed in England by A. Wheaton & Co., Ltd., Exeter

D
621·3841'7
PHI

For Diana, Hywel and Cenydd

Contents

Preface

When I was a student at Imperial College, Third Year Honours students had to write an essay and deliver a short lecture on any subject of their own choice. I elected to give a lecture on early radio detectors, my interest in this topic having been awakened by reading the Science Museum *Handbook of Radio Communication* published in 1934. This interest continued after graduation, and over the years I have built up a set of references and given lectures under the auspices of various societies and institutions. It was after such a lecture, given at the Institution of Electrical Engineers Annual Conference on the History of Electrical Engineering at Manchester in 1975, that Dr. Brian Bowers and Mr. Keith Geddes made the suggestion that I might care to write a booklet on the subject for the Science Museum. It is from this suggestion that the present book has grown.

I am very grateful to the Director of the Science Museum for allowing me the privilege of being Visiting Research Fellow from February to September 1977, and to the Council of University College Swansea for granting me a period of sabbatical leave in order to take up this appointment. During this leave the burden of my laboratory duties and my tutorial classes was cast on my colleagues in the Department of Electrical and Electronic Engineering, and I am grateful to them for carrying this extra load.

Living away from home and travelling frequently to London is an expensive business, and I am indebted to the Science Museum for covering certain of the travelling costs, and to the Leverhulme Trust who awarded me a research grant to help defray expenses. My wife's fortitude in coping alone with the children in my absence was also a major factor in enabling this work to be undertaken.

It is a particular pleasure to acknowledge the assistance and kindness of the staff of the Science Library, South Kensington, who, with unfailing cheerfulness, helped me to chase up many of the obscure references containing the information on which this book is based. I am also very grateful to Keith Geddes, who read the manuscript and made numerous comments and suggestions for improvement.

I hope that other electrical engineers will find the subject as fascinating as I have, and that the rediscovery of some of the 'lost' phenomena described will remind them of the tremendous achievements of our founding fathers. The writer on the history of radio has one thing in common with the medieval historian: just as the lance is no longer used in warfare, having been rendered obsolete by later developments, so have many of these phenomena sunk into obscurity. Indeed, many of them have never been satisfactorily explained, and authors of the period often leave out of their accounts what we would nowadays consider to be vital information. One hazard, however, not shared with the medievalist is that it is quite possible for people who may have had first-hand experience of these matters to be with us today. If a mere stripling of forty odd years has got it wrong through lack of such experience, I crave their indulgence in advance.

Vivian J. Phillips   July 1978
Sketty Green
Swansea, Wales

Acknowledgments

Diagrams and photographs are reproduced by permission of the following:

The Royal Society, London (3.51, 5.30, 7.11)

The Royal Society, Edinburgh (5.32)

The Institute of Physics, London (7.6, 7.7, 7.9, 7.17)

The Science Museum, London (1.1, 1.2, 1.4, 3.4, 3.19, 3.28, 3.32, 4.6, 4.10, 5.9, 5.16, 5.21, 6.4, 10.3, 10.5, 10.6)

Taylor and Francis, Ltd, London (2.6)

Electrical Review, London (4.7)

Wireless World, London (4.5, 7.8, 7.18, 8.6)

McGraw-Hill Book Co. Ltd., England (6.3, 3.29, 3.39, 9.16, 9.17)

Charles Griffin and Company Ltd, London and High Wycombe (3.20)

Macmillan, London and Basingstoke (2.16)

Chapman and Hall Ltd, London (4.13, 4.14, 6.13a, 7.10, 7.14, 8.2, 9.3)

Longman Ltd, Harlow, England (3.23, 4.1, 7.3, 7.16, 8.10, 8.13, 9.14, 9.15)

British Patent diagrams by permission of the Controller, Her Majesty's Stationery Office.

At the time of going to press no copyright holders other than those listed are known.

Oliver J. Lodge 1851–1940

John Ambrose Fleming 1849–1945

Andrè Eugenè Blodel 1863–1938

Jagadis Chunder Bose 1858–1937

Michael Idvorsky Pupin 1858—1935   Gustave Ferrié 1868—1932

Ernest Wilson 1863—1932   Reginald Aubrey Fessenden 1866—1932

Gugliemo Marconi 1874—1937

William DuBois Duddell 1872—1917

Lee DeForest 1873—1961

Marquis Luigi Solari 1873—?

Elihu Thomson 1853—1937 *David Edward Hughes 1831—1900*

Heinrich Rudolph Hertz 1857—1894 *Edouard Branly 1844—1940*

Chapter 1

Introduction

Mankind has been endowed with the five senses of sight, hearing, touch, smell and taste. It is not surprising, therefore, that his scientific curiosity was aroused from the earliest times by the light he could see, the sounds he could hear and the heat he could feel. Although the physicist has rather neglected the senses of smell and taste, these too have received their share of attention from people with gastronomic interests and from those concerned with the manufacture of perfumes.

There are, however, many phenomena in nature which cannot be perceived directly, and these must be converted into a form appreciated by the senses before their presence is revealed. Thus a magnetic field must be rendered visible by the controlled swinging of a lodestone or the patterns of iron filings, and an electrostatic force must create visible sparks, audible cracklings or unpleasant shocks before we are aware of its presence. For this reason radio waves lay undiscovered until the nineteenth century although they had always been produced naturally during thunderstorms. During that century various scientific workers became aware that there was some unknown phenomenon which was causing inexplicable results in some of their experiments.[1] For example, in 1842 Professor Joseph Henry noticed that the discharge of a Leyden jar (the form of capacitor used in those days[*]) was able to magnetize needles situated in the basement thirty feet below although two fourteen-inch thick ceilings intervened.[2] He also noticed that lightning flashes seven or eight miles distant produced a similar effect. Again in 1875, Professor Elihu Thomson was using a sparking coil in a room on the first floor of the Central High School in Philadelphia. He

*The 'jar' was used as a unit of capacitance, and the Admiralty Handbooks continued to quote the conversion factor for jars to microfarads until well into the 1930s. J.A.Fleming stated in his book (Section 2, Reference 6) that the pint size jar had a capacity of about 1/700 μF and the gallon size 1/300 μF.

had connected one side of the spark-gap to earth and the other side to a large insulated metal can, and he discovered that when he held a sharp pencil point to a brass door-knob about a hundred feet away at the top of the building, small sparks could be seen each time the sparking coil operated in the room below.[3,4] In a paper written years later he recalled his experiences on this occasion and described an improved spark detection apparatus which was designed for him by Thomas Edison.[5] This consisted of two sharpened points forming a small spark gap contained in a dark box, which made it easier to observe the feeble secondary spark.

Perhaps the best documented of all these early observations were those made by Professor D.E. Hughes (who was actually a professor of music) in London between 1878 and 1880.[6-9] Professor Hughes had been experiencing difficulty in balancing an inductance bridge with which he was working (Fig. 1.1) and he eventually traced the trouble to a loose connection in the circuit. As it happened, he was also experimenting with some of his primitive microphones in the same laboratory. One of these microphones consisted of a steel needle lying in loose contact with a piece of coke, the whole being connected in series with a battery and a telephone earpiece. (Other types of loose-contact microphones such as those of Fig. 1.2 were also used by him.) Sound vibrations impinging on this microphone would vary the contact resistance and hence the current, thereby reproducing the sound in the earpiece. Hughes noticed that when the faulty contact on his bridge was made and broken this created noises in the earpiece even though the two sets of apparatus were far apart and unconnected in any way. He was quite excited by this discovery and went on to carry out many experiments,

Fig. 1.1 *Hughes's induction balance circuit*
[Science Museum photograph]

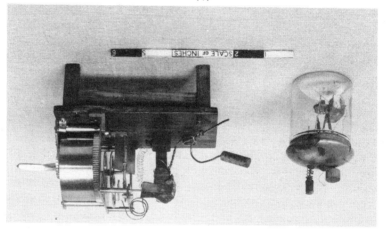

(a)

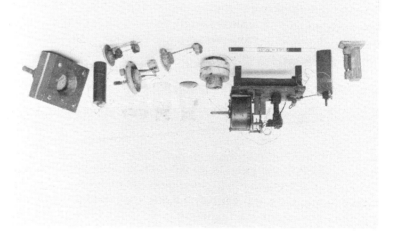

(b)

Fig. 1.2 *Hughes' microphone detectors*
    (a)   selection of detectors
    (b)   steel-needle/coke detector in glass jar. The clockwork device at
        the right (also seen in (a) above) was an automatic interrupter
        to allow experiments on reception at a distance without need for
        an assistant to operate the transmitter

including the reception of signals in the road outside his house in Great Portland Street. He showed his results to various people, and on 20th February 1880 he arranged a demonstration for the President and several Fellows of the Royal Society. In spite of the clear success of the

demonstration and the visitors' initial favourable reaction, they
eventually declared that the results could be explained by known
electric and magnetic phenomena and were in no way a proof of the
existence of the electromagnetic radiations which James Clerk Maxwell,
the Scottish physicist, had predicted in his theoretical studies of 1864.
The frontispiece shows the pages of Hughes' notebook in which he
recorded his experiences on this occasion. Such eminent scientists
having pooh-poohed his work, Hughes did not venture to publish it
until many years later, but his notebooks (kept in the British Museum)
illustrate the range of experiments which he carried out and establish
his claim to be one of the pioneers of radio science.

It was in the years 1886–88 that the German Heinrich Hertz put
the whole subject of electromagnetic radiations on a firm footing in a
series of classic experiments which established beyond all doubt that
these 'aetheric radiations', which we now call radio waves, most certainly
did exist and that they possessed properties similar to light waves in
that they could be reflected, refracted, polarized etc.[10] The transmitting apparatus which Hertz used is shown in Fig. 1.3. It consisted
of two small metal balls forming a spark-gap connected across the
terminals of a sparking coil (or 'inductorium' as it was sometimes
called). To each ball was attached a rod connected at the outer end to
a large metal plate or sphere. The whole thing formed a capacitor
which was charged up by the high voltage produced by the coil. When
this voltage became sufficiently high a spark occurred across the gap so
that the two plates of the capacitor were effectively connected together
via the rods, which possessed the property of inductance. It is now well
understood that in such circumstances the resulting current which flows
is oscillatory in nature, the charge surging to and fro between the plates.
In the early nineteenth century the possibility of such an oscillation
was the subject of much debate and dispute, but the experimental work
of Henry and others, coupled with the theoretical studies of William
Thomson (later Lord Kelvin) established beyond all doubt that oscilla-

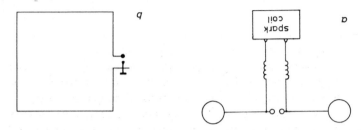

Fig. 1.3 *Apparatus used by Heinrich Hertz in his classic experiments of 1886–88*
(a) Spark transmitter, (b) spark-gap detector or 'ring resonator'

tions could occur.[11] In Hertz's apparatus the oscillating flow of charge
set up electric and magnetic fields which caused the emission of radio
waves. It is perhaps interesting to note in passing that the oscillation was
often at a frequency of 80–100 MHz. in what we would now call the
v.h.f. range, although this varied greatly with the dimensions of the
apparatus used. The experiments were carried out over distances of a
few metres only.

Hertz's radiation detector, often referred to in the literature simply
as his 'resonator', is also illustrated in Figs. 1.3 and 1.4. It consisted of
a square or circular ring of wire broken at one point to form a small
spark-gap. When it was subjected to the radiations from the transmitter
a minute spark could be observed at this gap. A micrometer screw
arrangement was usually provided so that the width of the gap could be
adjusted for optimum results. Some experiments subsequently performed by Sir J.J. Thomson and Mr. Peace at Cambridge showed that
in order to produce a visible spark between two metal balls, a potential
difference of at least 300 V was needed.[12] Thus it will be appreciated
that this spark-gap detector was a very insensitive device indeed and it
is a tremendous tribute to Hertz that even with this crude apparatus he
was able to demonstrate all the effects previously mentioned. There is
some evidence to suggest that Hertz also used another form of receiver
consisting of a spark-gap and rods identical to those at the transmitter,[13] but on the whole he seems to have preferred the ring form of
Fig. 1.3.

It was soon quite clear to all that before any further serious work
could be done on this subject it would be essential to find some better
means of detecting the presence of radio waves. Many physical phenomena, some long since forgotten, were pressed into service for this purpose and a great deal of ingenuity was shown in the very best 'string
and sealing wax' traditions of experimental science. This book is a
compendium of such techniques and its purpose is to tell the story of
the search for better detecting devices and to illustrate the inventiveness and skill of the founding fathers of radio science.

Before embarking on the story, it would perhaps be useful to say a
word or two in general about the development of radio. In the first
decade or so after the work of Hertz it was largely a matter of investi-
gating a new phenomenon in the laboratory. The verb 'to detect' is
defined in Collins' *New Gem Dictionary* as 'to find out or discover
the existence or presence or nature or identity' of something, and this
describes exactly the purpose of the experiments which were conducted at that time and the function of the various types of receiving
devices which were developed. Alternative names for these detectors

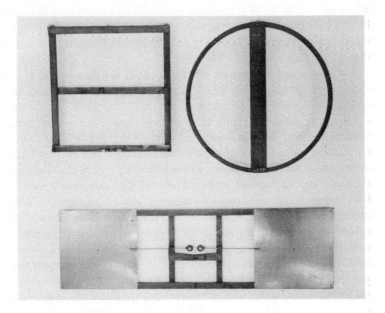

Fig. 1.4   *Experimental apparatus used by Heinrich Hertz in his work on radio waves (replica)*
[Science Museum photograph]

occasiónally encountered in the literature before the terminology became standardised are 'wave responders', 'revealers' and cymoscopes' (from the Greek *cyma*, a wave, and *skopein*, to see).

It was later realised that these new 'Hertzian waves' might well be useful for wire-less telegraphy, that is to say, for the transmission of Morse signals to distant places without the need for a wire connection. The reception of a Morse-type radio signal of dots and dashes separated by periods of silence really amounts to detecting whether or not a signal is present in the receiving aerial at any particular time, and so the word 'detector' was still very appropriate for the device which performed this function. Somewhat later, the techniques of amplitude-modulation were introduced, enabling speech to be transmitted directly by wireless telephony. The pieces of apparatus which recovered the speech signal from the radio carrier at the receiver were really the direct descendants of the wireless telegraphy 'detectors' and the word 'detection' continued in use to describe this operation until it was replaced by the better term 'demodulation' in quite recent times. The old word, though, is even now in the current vocabulary of the radio engineer, and he still refers to the 'envelope detector' and the 'ratio detector'. The radar engineer actually uses the word in a sense which is very near to the original meaning, implying discovering the existence of something.

Over the years, there were naturally developments on the transmitting side, too.[14] The original Hertzian transmitter generated isolated pulses of damped radio-frequency oscillation, these pulses occurring at more-or-less irregular intervals depending on the exact instants of sparking across the gap (Fig. 1.5a). If the detector happened to be a device which converted each pulse into a clicking sound in a telephone receiver the signal would be heard as an irregular buzzing sound. In a later development several spark-gaps were mounted on a rotating wheel which caused the sparks to occur at definite regular time intervals so that the received sound was a distinctive musical note which could be perceived more clearly in the presence of interfering signals. This was the so-called 'musical spark' form of transmission (Fig. 1.5b). There was a further advantage with this mode of operation. The spark-gap was connected in a primary circuit which was mutually coupled to a tuned secondary aerial circuit. During the spark, energy was transferred to the secondary. The motion of the electrodes quickly extinguished (or 'quenched') the spark, open circuiting the primary side. The transferred energy was then dissipated usefully in the aerial, none

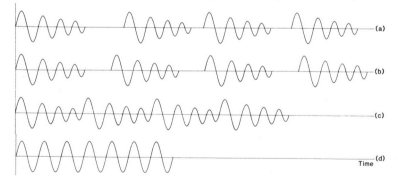

Fig. 1.5   *The transition from simple spark transmission to the continuous radio-frequency carrier (note that there would be many more cycles of r.f. oscillation within each burst than can conveniently be shown on a diagram of this sort)*
   (a)   Isolated irregular sparks
   (b)   Isolated sparks which occur at regular intervals; the 'musical spark' form of transmission
   (c)   Timed spark transmission. Here the phases of the separate bursts of r.f. are controlled so that they merge without gross discontinuities
   (d)   A continuous carrier

being returned to the primary circuit.

For various practical reasons it was very desirable to get away from this pulsed form of signal and to produce a continuous high-frequency carrier instead. If the individual bursts or pulses of Fig.1.5*b* were to be spaced more closely and with the correct phase, they could be made to run smoothly into each other producing an almost continuous signal such as that in Fig.1.5*c*. This was the 'timed spark' form of transmission A further development was the production of a truly continuous sinusoidal signal (Fig. 1.5*d*) either by generating it directly with a high-frequency alternator (producing long waves with frequencies in tens of kilohertz), or else by an arc transmitter which made use of the negative resistance of the carbon-arc to overcome the damping in a tuned circuit, thereby enabling it to oscillate continuously. Finally, the thermionic valve arrived on the scene to provide a convenient and easy method of generating a sinusoidal r.f. signal.

This book will cover a period extending from Hertz's experiments up to the coming of the crystal detector and the valve, which events may be considered as taking radio into the modern era. It must be appreciated, however, that even after the invention of the thermionic diode and triode (in 1904 and 1906–7) many other types of detector were still in use well into the Great War of 1914–18,[15] and it was not until the 1920s that the victory of the valve was complete.[16] Modern electrical engineers have often been brought up to believe that *rectification* is the basis of all radio detectors, but it is worth making it clear at the outset that most of the detectors to be described in this book were *not* rectifiers; they were instruments for indicating the presence or absence of a radio-frequency signal and, as we shall see, they used all sorts of varied phenomena. Rectification as a *means* of detection arrived at a comparatively late stage with the crystal and the diode.

Most of the detectors to be described were, by their nature, only able to provide a simple yes/no indication of the presence or absence of a received signal. A few were able to give a quantitative measure of the strength of that signal, but most of these were unable to operate with sufficient speed to follow the dots and dashes of a telegraphy signal. It will also be appreciated that they did not all perform their function equally well. Some were excellent, reliable and positive in action, while others were temperamental and (quoting one contemporary writer on the subject[17]) 'they occasionally exhibit very tricky antics'! Very few indeed progressed beyond the laboratory stage to see active service in practical communication systems. Those that were successful in this respect will be noted particularly, and some comparative figures will be considered in the final chapter.

References

1   FAHIE, J.J.: *Electrician*, **43**, 1899, p.204
2   TAYLOR, W.B.: 'A memorial of Joseph Henry' (Congress, Washington, 1880), p.256
3   SNYDER, M.B.: *General Electric Review*, **23**, 1920, p.208
4   THOMSON, E.: *Electrician*, **43**, 1899, p.167
5   THOMSON, E.: *General Electric Review*, **18**, 1915, p.316
6   HUGHES, D.E.: *Electrician*, **43**, 1899, p.40
7   CAMPBELL-SWINTON, A.A.: *JIEE*, **60**, 1921–22, pp.492
8   CAMPBELL-SWINTON, A.A.: *Times* 28th March 1922 (Letter)
9   *Daily Graphic*, 1st April 1922
10 DE TUNZELMANN, G.W.: *Electrician*, **21**, 1888, p.587 and following issues
11 THOMSON, W.: *Phil. Mag.*, Ser 4, **5**, 1853, p.393
12 THOMSON, J.J.: *Recent researches in electricity and magnetism* (Clarendon, 1893), p.89
13 BLAKE, G.G.: *History of radio telegraphy and telephony* (Chapman and Hall, 1928), p.54
14 O'DEA, W.T.: *Science museum handbook on radio communication* (HMSO, 1934) p.23ff
15 ROBINSON, S.S.: *Manual of wireless telegraphy for naval electricians* (US Navy, 1911), section 183
16 *Admiralty handbook of wireless telegraphy* (HMSO, 1920), p.264
17 EICHORN, G.: *Wireless telegraphy* (Griffin, 1906) p.81

Chapter 2

Improved spark-gap detectors

It was natural that the experimenters who followed Hertz and extended his researches should have tried first of all to improve the performance of his detector, since the spark which they were trying to observe was very feeble and difficult to see. Edison's pre-Hertzian attempt to improve Thomson's primitive pencil point has already been mentioned. Monsieur Joubert[1] constructed a detector of the form shown in Fig. 2.1*a*. This consisted of two rods held in a Y-shaped insulating block, forming a spark-gap at the centre, and carrying two flags of tinfoil at their outer extremities. One of the rods could be moved in and out by means of a screw so that the width of the spark-gap could be finely adjusted. A black cloth shade was mounted behind the gap to improve visibility, and the whole thing was of such dimensions that it could be picked up easily and brought close to the eye for careful scrutiny. It seems to have been important in these types of detector to set the width of the gap just right for best results, and other workers used micrometer screws in various ways in order to achieve this precise adjustment.[2] The detector described by Lodge in his book 'Modern views of Electricity' may be seen in Fig. 2.1*b*. Here the two rods were held in glass tubes mounted on a wooden framework. The ends of the rods were sharpened to form a spark gap, and the distance between the points could be set by rotation of the knurled wheel on the right.

Professor Augusto Righi of Bolgna constructed a special form of gap consisting of a small piece of silvered glass,[3-5] the silver coating being divided into two or more strips by fine cuts made with a diamond point, as in Fig. 2.2. The outer strips were connected in place of the normal two-ball spark gap. When radiated energy was received the

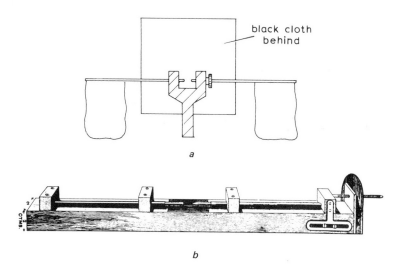

Fig. 2.1 Detectors having adjustable spark gaps
(*a*) Joubert
(*b*) Lodge
In both cases adjustment is by means of the wheel at the right
[Lodge, O.J. 'Modern views of Electricity', (Macmillan 1892) p.343]

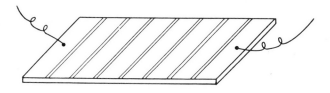

Fig. 2.2 *Righi's multiple spark-gap of silvered glass*

sparks now jumped from strip to strip and once the points of spark-ing had been located a microscope could be used to observe them in greater detail.

It was realised at quite an early stage that in order to achieve maximum sensitivity the receiving apparatus had to be 'syntonised' with the transmitter, or as we would say nowadays, they both had to be tuned to the same frequency. The two detectors shown in Fig. 2.3 were con-structed in such a way that their physical constants, and thus their resonant frequencies, could be varied either by sliding the telescopic rods in and out, or by moving the adjustable plates along the rods.[6, 7]

The position of maximum sensitivity was found by a simple process of trial and error. A rather more precise approach to tuning was that of Monsieur M.R. Blondlot,[8] who split the ring of a Hertzian detector and inserted two metal plates forming a capacitor, the dimensions of the plates having previously been carefully calculated to produce resonance at the desired frequency.

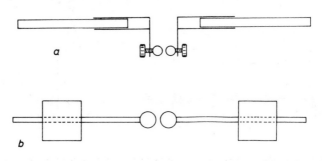

Fig. 2.3   *Adjustable spark-gaps to provide a degree of tuning or 'syntony'*
(a)   Fleming, (b)   Gregory

Even when all these refinements had been incorporated, the spark-gap was still a detector which could only be viewed conveniently by one person at a time. Many experimenters wished to demonstrate their apparatus to a larger audience in the lecture theatre and it is interesting to look at some of the ways in which they did this. A popular method was to connect a Geissler tube in place of the spark-gap.[9-12] This was a discharge tube containing gas at low pressure which glowed when a voltage was applied between its electrodes. Thus the spark was replaced by a flash in the tube, which was much easier to see. An extra refinement was to add a little fluorescent material inside the tube to brighten its glow still further.[13] Several people improved on this idea by using it in the form of a trigger tube.[14-18] Various shapes of tube and arrangements of electrodes were used, but the version devised by Zehnder, illustrated in Fig. 2.4, may be taken as typical. A partly evacuated tube containing air or some other gas at low pressure contained two main electrodes which were connected to a battery. The voltage of the battery was just below that which was necessary to make the tube strike and glow. The tube also contained a set of subsidiary electrodes, very close together, and connected in place of the normal spark-gap. Because of the close spacing, the incoming oscillations produced high electric fields in the gap, resulting in some ionisation of the gas. These ions then seeded the ionisation process in the whole tube, causing it to light up. This was a very effective method of demonstration since the brightness of the final display was dependent on the direct voltage from the battery and not on the feeble oscillatory voltage, which merely acted as a trigger.

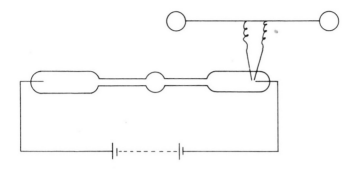

Fig. 2.4 *Zehnder's trigger-tube arrangement shown connected to a Hertzian receiver in place of the usual two-ball spark gap*

R.A. Fessenden,[19] as late as 1908, suggested the use of a tube of the form shown in Fig. 2.5. The idea here was that the voltage between points a and b would normally be kept just greater than that between a and c so that the glow would reside in the upper limb of the tube. The incoming oscillatory voltage was to be added to voltage $a-c$ so that the glow would transfer to the other limb, thereby demonstrating the arrival of a signal.

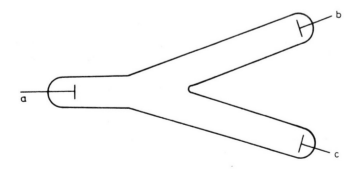

Fig. 2.5 *Fessenden's discharge tube*
This was connected so that the arrival of a pulse of r.f. transferred the glow from one limb to the other

Another approach to the demonstration problem was to allow the small spark to ignite an explosive mixture of gases.[20-21] Fig. 2.6 shows the apparatus constructed for this purpose by Lucas and Garrett. Here the two arms of the Hertzian receiver rest on the blocks (B and B′) so that the fine wires (P and P′) make contact with them. These wires form a spark-gap (G), the spacing of which can be adjusted by a nut (H) at the bottom. The spark-gap lies inside a glass tube which is partly filled with dilute hydrochloric acid. The two wires (E) at the bottom of the tube are connected to a battery so that hydrogen and oxygen are given off and collect in the upper part of the tube. The mixture is ignited by the small spark at G so that a loud pop is produced, clearly audible to all in the lecture room. Another similar method was to use an Abel's fuse connected across the spark-gap.[22] Sir Frederick Augustus Abel was Professor at the Woolwich Arsenal and his fuse was an electrical detonator consisting of a very fine wire, usually made of copper phosphide or a silver/platinum alloy, surrounded by explosive powder. A direct current just insufficient to set it off was allowed to flow and the extra oscillatory current from the resonator was then enough to explode the powder, giving a highly audible indication of the presence of oscillations!

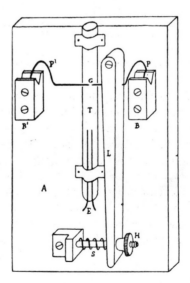

Fig. 2.6   *Arrangement of Lucas and Garrett whereby the feeble secondary spark ignited an explosive mixture of hydrogen and oxygen*
[*Phil. Mag.,* **33**, 1892, p.300]

A technique which sounds much less dangerous was used by L. Boltzmann [23], [24] and is illustrated in Fig. 2.7. It is based on a gold-leaf electroscope. This consists of a conducting rod fitted into a glass bottle through an insulating stopper and supporting two pieces of fine gold leaf at its lower end. If the rod is electrically charged, the like charges on the leaves repel and they are forced apart to the positions shown in the Figure. He used this in conjuction with a spark-gap, charging it up until it was almost sparking over to earth. The radio-frequency voltage caused by the incoming radiation was also applied across the gap, the combined voltages being then sufficient to cause a spark to pass. The charge then flowed away to earth so that the leaves collapsed, giving a visible indication that the spark had occurred.

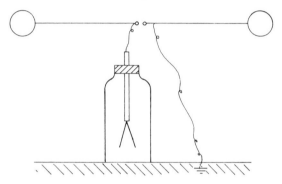

Fig. 2.7   *Boltzmann's gold-leaf electroscope arrangement*

G.F. Fitzgerald[25 – 26] tackled the problem by using an ordinary moving-coil galvanometer, as shown in Fig. 2.8. When a spark occurred across the main gap of the Hertzian receiver some charge was also found to flow across the smaller gap and through the galvanometer, which caused its needle to give a 'kick', rendering the effect visible to a large audience. E.J. Dragoumis at Liverpool and R.A. Fessenden in the USA suggested using a piece of paper soaked in potassium iodide solution and in contact with two wires connected to each side of the spark-gap.[27 – 28] Under the action of the current which flowed, the iodide solution discoloured, producing a permanent record of the reception of the radio wave. Albert Turpain[29 – 30] used the arrangement shown in Fig. 2.9. The ring detector was split and a battery and telephone earpiece were connected in series across the gap thus formed. Normally no current was able to flow because of the main gap, but when this was rendered conducting by the feeble spark, battery current was allowed to flow momentarily and a click was heard in the earpiece. A galvano-

meter could also be used in place of the earpiece, if preferred, for
lecture demonstrations.

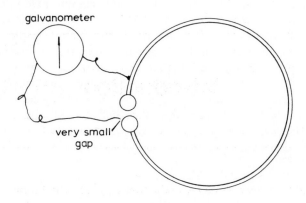

Fig. 2.8   *Galvanometer circuit used by Fitzgerald*

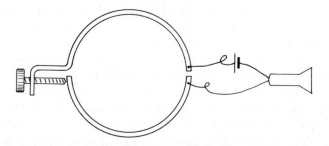

Fig. 2.9   *Use of a telephone earpiece with a spark gap to provide an audible
indication (described by Turpain)*

All these refinements and additions, useful and ingenious as they undoubtedly were, still depended basically on the same insensitive spark-
gap. In the search for improved performance many other types of
detector were invented, and these will be described in succeeding
chapters. In order to bring some sort of order into the story an attempt
will be made to classify them according to the types of physical effects
they employed. As always in such cases, some do not fit obligingly into
whatever categories may be chosen, and some could be considered in
more than one chapter. Such cases will be noted in the text.

References

1 *Electrician*, **23**, 1889, p.452
2 TURPAIN, A.: *Les applications pratiques des ondes electriques* (Carré et Naud, Paris, 1902), p.17
3 Reference 2, p.32
4 SEWALL, C.H.: *Wireless telegraphy* (Crosby-Lockwood, 1904), p.155
5 BLAKE, G.G. *History of radio telegraphy and telephony* (Chapman and Hall, 1928) p.57
6 FLEMING, J.A.: *Principles of electric wave telegraphy and telephony (Longmans, Green, 3rd Edn., 1916) p.466*
7 GREGORY, W.G.: *Proc. Roy. Soc. Lond.*, **10**, 1889, p.290
8 BLONDLOT, M.R.: *Journal de Physique*, **10**, Ser. 2, 1891, p.549
9 DRAGOUMIS, E.J.: *Nature*, **39**, 1889, p.548
10 BLONDEL, A., and FERRIÉ, G.: *Electrician*, **46**, 1900, p.21
11 Reference 2, p.30
12 RIGHI, A.: *L'Ottica delle oscillazioni elettriche* (Zanichelli, Bologna, 1897), p.13
13 MOORE, B.E.: *Phys. Rev.*, **4**, 1896, p.149
14 ZEHNDER,L.: *Electrician*, **30**, 1892, p.253
15 Reference 6, p.529
16 Reference 2, p.30
17 HARDEN, J.: *Review Électrique*, **3**, 1905, p.185
18 BOUASSE, H.: *Oscillations Électrique* (Delagrave, Paris, 1921), p.321
19 FESSENDEN, R.A.: British Patent 2685, 1908
20 LODGE, O.J.: *Electrician*, **40**, 1897, p.89
21 LUCAS, W. and GARRETT, T.A.: *Phil. Mag.*, **33**, 1892, p.299
22 ABEL, F.A.: *JIEE*, **3**, 1874, p.268
23 BOLTZMANN, L.: *Wied. Ann. Phys. und Chem.*, **40**, 1890, p.399
24 DE TUNZELMANN, G.W.: *Wireless telegraphy – a popular explanation* (Office of Knowledge, London, 1901), Chap. 5
25 FITZGERALD, G.F.: *Nature*, **42**, 1890, p.173
26 FITZGERALD, G.F.: *Wireless telegraphy*. Royal Institution Library of Science, Vol. 4 (Elsevier, 1970)
27 Reference 9
28 FESSENDEN, R.A.: US Patent 706743, 1902
29 Reference 2, p.19
30 *Science Abstracts*, **2**, 1899, No. 691, p.292

Chapter 3

Coherers

The coherer was perhaps the most important of all the early radio-wave detectors, and it was used in many different forms. It made use of a phenomenon which occurs in a poor electrical contact, the sort of contact which the engineer of today would call a 'dry joint'. Such an imperfect contact between two conductors normally exhibits a very high electrical resistance due, in large part, to the thin film of oxide which exists between the two metals. When an alternating or direct voltage is applied between the conductors this resistance decreases quite markedly. A voltage of a few tenths of a volt is often quite sufficient to produce the effect. To give some idea of the magnitudes of the resistance changes involved we may quote the results of an experiment carried out by Professor Edouard Branly, a pioneer in the application of this effect to radio reception.[1] Two oxidised copper rods lying in loose contact showed between them a resistance of 80 000 Ω. After the application of a voltage this fell to 7 Ω. Another early experimenter named Von Lang[2] reported a change from infinity to 380 Ω. Many poor contacts exist in a loosely packed mass of metal filings, and these also exhibit a similar drop in resistance on application of a voltage.

This phenomenon of 'coherence' seems to have been noticed first by the Swedish scientist P.S. Munk af Rosenschöld of the University of Lund[3-6] as early as 1835, and it was rediscovered in 1852 by S.A. Varley, who applied it to the protection of overhead telegraph wires from lightning strikes.[7] Carbon granules were loosely packed into a wooden box having two electrodes inserted from opposite sides, and this was then connected between the overhead wire and earth. With the low voltages normally used for signalling the comparatively high resistance of the mass of carbon granules had no effect, but in the event of a high voltage appearing during a lightning surge the resistance dropped sharply so that the surge current was diverted to earth and the coils of the receiving apparatus were protected from damage.

The effect in metal filings was rediscovered yet again in 1884 by Professor Temistocle Calzecchi-Onesti,[8-10] and the effectiveness of radio-frequency voltages in producing coherence was noted by Branly in 1981.[11-15] Professor Hughes' pre-Hertzian detector which has already been described was clearly of the coherer type, since his microphones consisted essentially of poor contacts, the resistances of which were affected by sound vibrations. For this reason, radio-wave detectors of the coherer type were sometimes called 'microphonic' detectors. Professor J. Chunder Bose was of the opinion that 'molecular receivers' would have been a better name for them.[16] while Edouard Branly himself rather favoured the term 'radio conductor'[17] The German word 'fritter' was also suggested,[18] but it seems to have been Oliver Lodge who coined the word 'coherer', and it is by this name that they have been known ever since in the English-speaking world.[19-20]

During the period when coherers were being used there was considerable discussion and disagreement as to how exactly this resistance change came about. In fact, the phenomenon was never satisfactorily explained at that time because other, better, devices superseded them and interest was lost before the matter was resolved. We shall return to this fundamental question later, but for the moment we shall simply look at some of the devices in which the phenomenon of coherence was used.

One point needs to be clarified before we proceed further. With almost all these devices the low-resistance condition persists after the coherence has taken place. Once a signal has been received, the device coheres, and unless something is done about it, it is then unable to respond to the arrival of further signals. It can, however, be restored to the high-resistance (sensitive) condition by mechanically shaking or tapping it, and practical systems using coherers incorporated some means of restoring the sensitivity in this way in order to prepare for the reception of further signals. The various methods of effecting this restoration will be considered later, but first we shall examine the coherers themselves in some detail. It will be convenient to consider first those devices in which only one, or very few, poor contacts were used, and to look at the multicontact filings coherers afterwards.

The coherer itself could be connected in place of the spark gap of a Hertzian ring-detector, as shown in Fig. 3.1. In this arrangement, which was used by Lodge and Muirhead,[21] the cohering contact (c) would normally have a high resistance so that very little current would flow from the battery (b). When a radio wave was received, the oscillatory voltage induced in the loop would cause the resistance to fall so that a

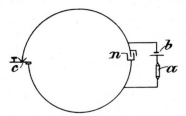

Fig. 3.1   *A simple single-contact coherer used by Lodge and Muirhead in place of the spark-gap in a Hertzian 'ring resonator'*
The capacitor *n* provides for some degree of tuning; *b* = battery, *a* = galvanometer
[British Patent 18 644, 1897]

current would flow, being indicated by a galvanometer of some sort (a). The capacitor (n) was for purposes of tuning or 'syntony'. The similarity between this and the spark-gap detector of Fig. 2.9 is very marked. The need for restoration was really the only difference.

At a somewhat later stage the coherer was usually connected between an aerial and earth, as shown in the basic operating circuit of Fig. 3.2. When the coherer was in the high-resistance condition (often

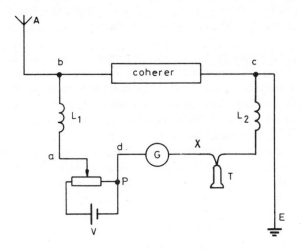

Fig. 3.2   *Basic coherer operating circuit*
A = aerial, E = earth, G = galvanometer, T = telephone receiver, V = battery

simply referred to as being in the 'sensitive' state) little battery current would flow around the circuit *abcd*. An oscillating voltage appearing between aerial and earth would cause coherence so that appreciable battery current was then able to flow around the loop. This sudden increase could be heard as a click in the telephone earpiece (T) or be seen as a deflection of the galvanometer (G). Some coherers were subject to a threshold effect[22] and their sensitivity was much improved by the application of a small direct voltage across them. For this reason the potentiometer (P) was included so that the 'bias' voltage could be adjusted for best results. The battery (V) thus performed two functions in this circuit: (i) it supplied a bias voltage to bring the coherer to the condition of maximum sensitivity, and (ii) in conjunction with the galvanometer and telephone it provided a means of sensing when the change to low resistance had occurred. The two choke coils L_1 and L_2 were included so that the oscillatory current from the aerial would not be shorted out through the battery circuit, reducing the voltage appearing across the coherer itself. They were also required for another purpose which will become apparent later.

Coherers took many different forms – almost as many as the number of people who used them, in fact. Let us look at some typical examples. The one illustrated in Figs. 3.3 and 3.4a was used by Branly and consisted of a small steel tripod with pointed and slightly oxidised steel legs resting on a plate of polished steel, the cohering resistance being that between the tripod and the plate.[23-26] An alternative version[27] of the tripod coherer had legs terminated with gold telluride points resting on a plate of polished silver, the whole being contained in an evacuated enclosure. It is not clear from the literature whether the advantages of this expensive version were real, or merely psychological!

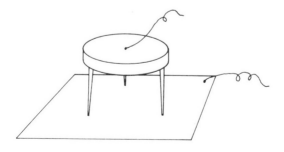

Fig. 3.3   *Branly's tripod coherer*

Fig. 3.4   *Various single-point coherers*
  (*a*)   Branly's tripod
  (*b*)   Lodge's spiral coherer
  (*c*)   Lodge's flat-spring coherer
  [Science Museum photographs]

A version of the coherer due to F.L. Odenbach[28-29] is shown in Fig. 3.5. Four brass posts, mounted on a base of cherry-wood supported two slings made of German silver or steel wire, and a light graphite rod rested on the two wires. The baseboard was 2 in x 3 in,

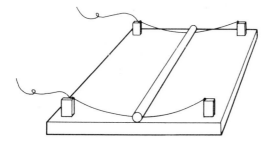

Fig. 3.5 *Odenbach's coherer*
The rod is suspended on the two wire loops

and the inventor reported that A.W. Faber's Siberian leads for artists' pencil no. 5900 gave the most perfect results! Jagadindu Ray[30] made use of the configuration of Fig. 3.6, where several rings made of iron or steel wire were held between two bar-magnets mounted in an insulating block, while C. Schniewindt[31,32] constructed a loose wire gauze with a spiral or zigzag cut made across it, as in Fig. 3.7. Current passing from one end to the other had to pass through many imperfect contacts, each of which exhibited coherence on the arrival of a radio-frequency oscillation.

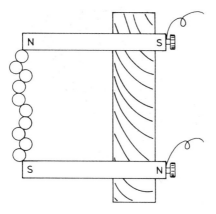

Fig. 3.6 *Ray's coherer*
The iron or steel rings form a loose chain between the magnetic poles

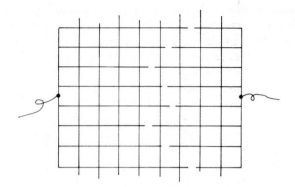

Fig. 3.7   *Schniewindt's coherer consisting of a loose gauze of wire with cuts in places*

A rather neater arrangement was Maskeleyne's 'conjunctor'[33–35] (Fig. 3.8). This consisted of an oxidised steel cylinder (11) with bevelled ends resting on two polished steel hemispheres, the whole being contained in a tubular glass envelope and having a general appearance similar to that of a modern cartridge fuse. One of the hemispheres had a small cavity cut into it, and some anhydrous calcium chloride was put into this to keep the air dry inside the glass envelope. This was quite a carefully engineered device, and its inventor clearly regarded it as being a cut above other people's coherers. As he explained in his patent specification, he used the word 'conjunctor' for it because this 'sets it

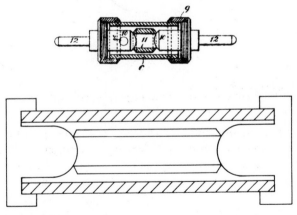

Fig. 3.8   *Maskeleyne's 'conjunctor' — a carefully engineered coherer; the lower diagram is simplified.*
[British Patent 16 113, 1903]

apart from irregular and heterogeneous point-and-surface contacts known as coherers. All components are capable of exact reproduction in any quantity, and the accuracy is reflected in its extremely regular action in practical use'. He also provided very precise instructions as to how the oxide coating was to be produced on the steel cylinder.

All these coherers which have been described so far undoubtedly worked, as the results reported by their inventors testify, but with the possible exception of the conjunctor they must have been very awkward to set up as no form of adjustment was provided. This defect would have been very obvious to anyone working with them, and some of the methods used to provide such adjustment will now be described. A very crude method was used by Bowlker, whose coherer consisted of one strip of metal laid horizontally across two others.[36] The contact pressure between them was altered by the simple expedient of putting weights on the top strip. Oliver Lodge used the arrangement of Fig. 3.9 and 3.4b.[37–38] A wooden case contained a thin spiral spring of steel

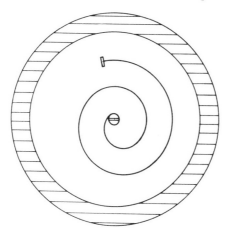

Fig. 3.9   *Lodge's spiral coherer*
The end of the spiral wire bears lightly on the small aluminium plate

fixed to a shaft at the centre. The free end of the spring pressed lightly on a small aluminium plate, the merit of this arrangement being that the contact pressure could be adjusted by rotation of the central shaft. Another of Lodge's adjustable coherers[39–40] is shown in Figs. 3.10 and 3.4c. This made use of a flat spring, and contact between this and a metal point could be adjusted by turning the screw on which the point was mounted, or by another screw bearing on the end of the spring. Quite a popular form of adjustable coherer used several steel balls in a glass tube, as in Fig. 3.11.[41-44] Some form of spring was used to make contact with the end balls, and the contact pressure between the balls could be varied by tilting the tube at an angle to the horizontal. Axel Orling and G.G. Braunerhjelm constructed an elaborate form of gimbal mounting, shown in Fig. 3.12, to ensure that, once correctly set, the tube would be maintained at the correct inclination even if the receiver were aboard ship.[45] A still more elaborate version patented by the same two inventors may be seen in Fig. 3.13.[46-48] Here the steel balls or cylinders are arranged in two tiers (a and a') within an evacuated enclosure (O). Rotation of the horseshoe magnet (MM') about its axle (r) turns the armature (A) situated inside the enclosure. This moves the contact (k^2) in and out which causes the upper rods to ride up over the lower ones to a greater or lesser extent so that the contact pressure between them will vary.

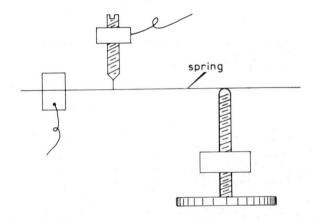

spring

Fig. 3.10   *Lodge's adjustable flat-spring coherer*

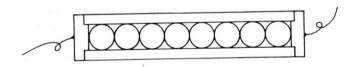

Fig. 3.11   *Steel ball coherer used by Branly and others*

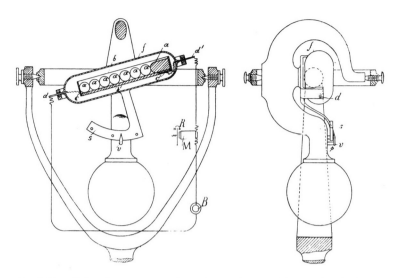

Fig. 3.12 *Ball coherer mounted in gimbals for use aboard ship (Orling and Braunerhjelm)*
[British Patent 1866, 1899]

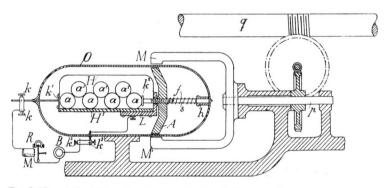

Fig. 3.13 *Adjustable ball coherer devised by Orling and Braunerhjelm*
[British Patent 1867, 1899]

Professor Chunder Bose[49-50] made extensive use of the contacts between spiral springs, one of his arrangements being shown in Fig. 3.14. The spiral springs are here resting in a U-shaped cavity cut into a block of ebonite, and the contact pressure between them may be adjusted by means of the knurled knob which causes a metal plate to move against them. Using a spiral-spring coherer such as this in conjunction with an Abel's fuse (see page 14) inserted into the touchhole he was able to demonstrate the radio-controlled firing of a cannon.[51]

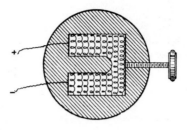

Fig. 3.14 *Adjustable spiral-spring coherer used by J. Chunder Bose*
[Electrician, 36, 1895, p.291]

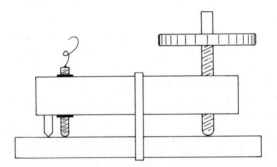

Fig. 3.15 *Adjustable single-contact coherer used by Taylor for precise measurement of the coherer effect*

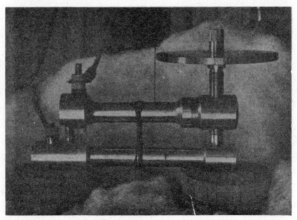

Fig. 3.16 *Photograph of Taylor's coherer*
[Physical Review, 16, 1903]

Several workers attempted to make detailed measurements of the properties of a cohering contact by using more refined methods of pressure adjustment. Possibly the most sophisticated experiments were carried out by A.H. Taylor[52-53] who, in 1903, published an extensive set of results which he had obtained with the arrangement shown in diagrammatic form in Fig. 3.15. The coherer itself is also shown in Fig. 3.16. Two steel bars seven inches in length are pivoted on a knife edge at one end, being held apart at the other by a micrometer screw arrangement. The whole is held together by a rubber band passing around the two bars. A metal point is mounted on the top bar in an insulating bush, and it is the contact between this and the lower bar which is under observation. Bearing in mind the leverage effect of the bars, extremely fine control of contact pressure may be obtained in such an arrangement, and in fact Taylor claimed to be able to set the contact spacing to within 1/68 000 of an inch, although he admitted that the adjustment in practice was not quite as good as this on account of some slight play in the micrometer. With such a sensitive piece of apparatus it was naturally very difficult to obtain reliable and repeatable results, the slightest mechanical vibration being sufficient to disturb it. Taylor's own words illustrate very graphically the difficulties he encountered:

I have found it absolutely impossible to get any satisfactory results in the daytime, and the data to which I attach most importance were obtained after midnight when the electric cars had ceased running, and the campus was quiet. It was always found necessary to stop the clocks and shut off the steam in order to avoid the 'thumping' in the pipes. Many a set of observations has been abruptly terminated by the passage of a waggon on a road fully two blocks away.

Of course the observer must use the utmost precaution against mechanical jars. His chair must not squeak, and if it is necessary to repeat the observations to a second person, it had best be done in a whisper, as the rough guttural tones often affect the coherer if the speaker is close to it.

The present author has carried out measurements on coherers in his own laboratory and can certainly endorse these heart-felt comments.

Another experimenter who made careful measurements was P.E. Robinson,[54] and the apparatus he used is shown in Fig. 3.17. In this apparatus two stout bars with hemispherical ends are suspended by wires from two blocks (C_1 and C_2) which rest on the horizontal member of a wooden framework. Contact is made to the bars via wires dipping into two cups of mercury. By movement of the blocks along the frame fine control of the contact pressure may be achieved.

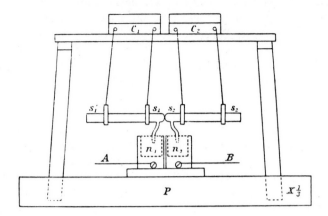

Fig. 3.17   *Robinson's apparatus for investigation of the phenomenon of coherence*
[*Ann. der Physik,* Leipsig, **11**, 1903, p.755]

The coherers described thus far used few imperfect contacts, but the most successful and most widely used coherers were those which contained metal filings. These multicontact devices again appeared in many different forms and a selection is shown in Fig. 3.18–3.20. Filings of all sorts of metals were tried, but those which were relatively easy to oxidise, such as copper, nickel and iron, were found to be most satisfactory. Some workers claimed success with the noble metals silver, platinum and gold, but on the whole these seem to have been more temperamental to operate.[55–61] It is worth noting here that filings of a few metals such as potassium and arsenic could, under certain circumstances, exhibit behaviour opposite to that expected in that their resistance *increased* under the influence of the applied voltage.[62,63] Not unnaturally, devices using such metals were termed 'anticoherers', and indeed this name came to be applied generally to any detector, whatever its physical basis, whose resistance increased in this way. Some of these will be described later in this book.

In Branly's filings coherer (Fig. 3.18a) the filings themselves were held between metal plugs inserted into an insulating tube. Lodge[64] seems to have preferred the rather longer tube of Fig. 3.19a containing very coarse particles such as metal turnings from a lathe. He also used the structure of Fig. 3.18c, where the filings were contained in a sealed glass envelope, two wires inserted through the glass acting as contacts.[65] Popov (also spelt Popoff), the noted Russian pioneer, used a similar arrangement to that of Branly (Fig. 3.18b), save that the con-

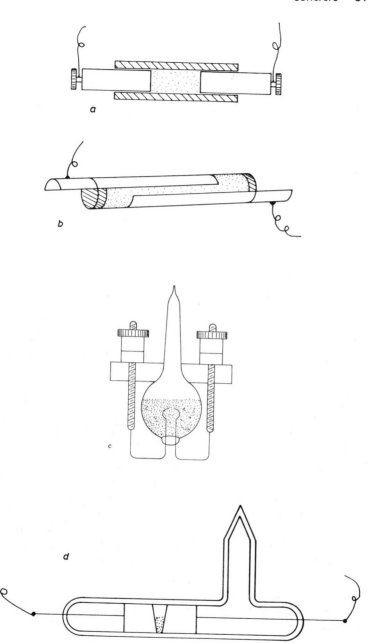

Fig. 3.18   *Various forms of filings coherer*
(a) Branly, (b) Popov, (c) Lodge, (d) Marconi

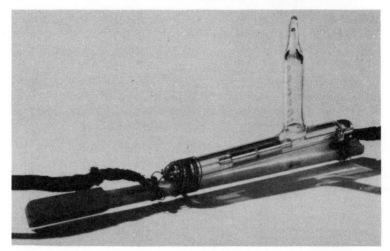

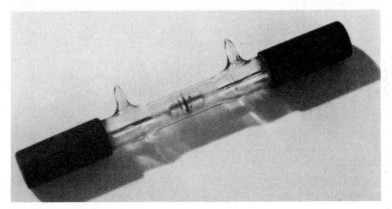

Fig. 3.19   *Various filings coherers*
(*a*)   Lodge,   (*b*) Marconi,   (*c*) Slaby
[Science Museum photographs]

Fig. 3.20 *Set of Telefunken filings coherers in their case*
[Eichorn, G.: *Wireless telegraphy* (Griffin, 1906) p.80]

tacts were two metal plates inserted along the sides of the tube.[66] He used this coherer in 1893 to record the occurrence of distant lightning flashes, which, being electrical spark discharges, produce pulses of radio waves. Initially he used a receiver only, and it was rather later that he used a man-made transmitter to signal at a distance.

Quite a neat coherer was that produced by W.W. Massie in 1905 (Fig. 3.21).[67-68] In this version the filings were contained in a metal cup (11) with a silver lining (13). A glass tube (9) supported an insulated cap (10) through which a steel needle (14) dipped down into the filings. It was suggested that the filings should be a mixture of magnetic and nonmagnetic metals, and that the steel needle should be magnetised. The magnetic particles would thus stick to the end of the needle, forming a 'tree', so that the contact area with the other particles would be increased, as shown in the small diagram of Fig. 3.21.

One of the most refined coherers was that developed by Marconi[69] illustrated in Figs. 3.18d and 3.19b. He made the overall dimensions of the tube much smaller than in those previously mentioned (about 4 cm long and 0·5 cm wide) and he also made the gap slightly wedge-shaped.

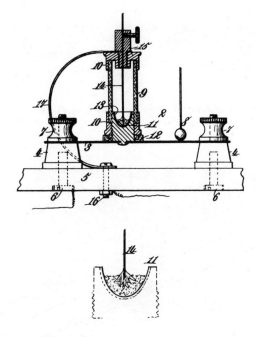

Fig. 3.21 *Massie's filings coherer,*
The needle (14) which dips into the filings is lightly magnetised. The coherer is mounted on a springy strip (3) and is decohered by a blow from the tapper (8)
[US Patent 800 119, 1905]

This feature, which had also been used by Camille Tissot[70] and Adolf Slaby (Fig. 3.19c) meant that by rotating the tube about its long axis the filings could be made to lie in a wider or narrower gap, thus providing some degree of sensitivity control. Another coherer having an adjustable gap, made by Ducretet of Paris, a firm of instrument manufacturers, is shown in Fig. 3.22. Marconi used filings consisting of 95% nickel mixed with 5% silver lying between highly polished silver plugs, the ends of which had been amalgamated with mercury. The coherer tube was evacuated of air in order to prevent eventual excessive oxidation of the particles, thereby stabilising the long-term performance of the coherer.

The experience of a new recruit to the Marconi establishment makes interesting reading:[71]

> Dowsett hung around feeling silly while everyone went on working. Finally Marconi handed him a piece of metal no larger than a shilling and the oldest and smoothest of files, telling him to make some metal filings for a coherer. For half an hour he filed away

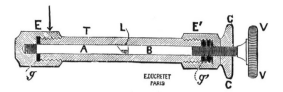

Fig. 3.22   *Coherer with gap of adjustable width made by Ducretet of Paris*
[Mazzotto, D.: *Wireless telegraphy and telephony* (Whittaker, 1906),
p. 174]

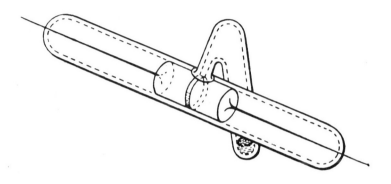

Fig. 3.23   *Blondel's side-pocket coherer*

[Fleming, J.A. *'Principles of electric wave telegraphy and telephony'*
(Longmans, 3rd Edn., 1916) p.479]

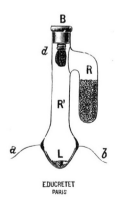

Fig. 3.24   *Side-pocket coherer by Ducretet of Paris*
[Mazzotto, D.: *Wireless telegraphy and telephony* (Whittaker, 1906),
p. 174]

furiously, increasingly convinced that the whole thing was a leg-pull for the 'new boy'. He accumulated a minute heap of dust, out of all proportion to the effort he had expended and was surprised when Marconi told him he had enough to fill a coherer. Only a fine, clogged-up file, it seemed, produced particles minute enough to serve this purpose.

Another method of varying the sensitivity was to have a gap of constant width, but to vary the amount of filings lying in that gap. To this end Blondel introduced his side-pocket coherer (Fig. 3.23).[72-75] The glass pocket at the side of the main tube contained filings so that by appropriate juggling the required quantity could be transferred to the gap. Other coherers which varied the sensitivity in this way are shown in Figs. 3.24 and 3.25. In Ducretet's version the filings lay between two wires inserted into the bottom of a glass tube and the filings reservoir was sealed onto the upper part of the tube.[76] The small bag (*d*) contained a substance to dry the air. Ferrié's version had a channel H cut into one of the metal plugs, and this communicated with the main gap by a small groove cut along the top of the plug.[77-78] Once again, by appropriate manipulation, filings could be transferred between gap and reservoir. It is opportune to point out here that great sensitivity was not always desirable, and in some cases could be a positive nuisance.[79] If the level of atmospheric electrical activity happened to be high a very sensitive coherer could be maintained in an almost permanent low-resistance state, preventing reception of the wanted signal. In such cases it was much better to use a less sensitive tube which would be unaffected by atmospherics.

Fig. 3.25   *Ferrié's arrangement for adjusting the quantity of filings in the gap*
By appropriate manipulation, filings could be transferred from the channel H along groove *r* into the main gap I
[Mazzotto, D.: *Wireless telegraphy and telephony* (Whittaker, 1906), p. 173]

One other factor which determined the sensitivity was the size of the filings.[80-81] With this in mind, S.R. Bottone suggested the set-up illustrated in Fig. 3.26, which he called his 'multilocular polytonic coherer', consisting of seven separate coherer tubes connected in parallel across the triangular metal framework as shown.[82] The individual coherers were constructed as shown in the lower diagram, and each contained a different grade of filings. Bottone was quite insistent that the tube containing the finest filings should be at the narrow end of the framework, but it is by no means clear why this should be so. Neither is it clear that the operation of coherers in parallel in this way would be of any great advantage. Presumably the idea was that the optimum grade of filings would be operational in any given conditions, but the shunting effect of all the others must surely have been a disadvantage. However, in fairness it must be said that Albert Turpain[83] also connected coherers in parallel and since from the general tone of his book Bottone seems to have been quite a practical man one is inclined to think that it must have been beneficial in some way.

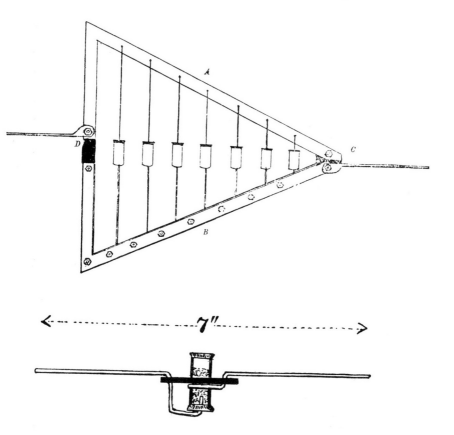

Fig. 3.26   *Bottone's 'multilocular polytonic' coherer*
The lower diagram shows the construction of each individual coherer
[Bottone, S.R.: *Wireless telegraphy and Hertzian waves* (Whittaker, 1910), p.105 and 108]

Restoration of coherers

It was remarked earlier that the drop in resistance on coherence persists until the cohering contacts are shaken or mechanically disturbed.
Thus when a signal has been received it is necessary to tap the coherer
back to the high-resistance state to prepare it for the next one. In the
jargon of the day this was known as 'decohering' or 'restoration' and
it is interesting to look at some of the ways in which it was achieved.
Perhaps the best remembered is the so-called 'coherer and tapper'
method of operation, widely used by Marconi, Popov, Slaby and
others.[84-88] Referring back to the general circuit of Fig. 3.2, a bell-
trembler mechanism was included in the battery circuit at the point x.
When a signal was received the resulting battery current caused the
trembler to operate, and the hammer of the bell was made to beat
against the coherer tube, creating the mechanical shock necessary to
restore it to the high-resistance condition. Actually, it was more satisfactory in practice to include a relay at the point x and to make this
operate the bell by using a completely different battery. The additional
function of the two chokes L_1 and L_2 can now be explained. When the
bell trembler is operating, the small spark at its make and break points
can cause radio-frequency oscillations to be generated. These travel
back along the wires, tending to keep the coherer in a permanent state
of coherence and making it difficult to restore it to the sensitive condition. Much more reliable action is ensured if the chokes are present to
prevent this. The coils seem to have been introduced first by Popov, but
they soon became a fairly standard feature of the coherer and tapper
circuit.[89]
    Another way to avoid this complication was to do away with the
trembler mechanism and to rely on a single tap with the hammer to
produce the desired decoherence. A circuit used by Branly for this purpose is shown in Fig. 3.27.[90] The flow of current in the solenoid (H')
attracts the armature (N) and causes the hammer (M) to strike the
coherer. A further solenoid (H) operates a pen B which makes a permanent record of the received signals on a moving paper tape. Some examples of coherer and tapper arrangements may be seen in Fig. 3.28
and 3.29. Fig. 3.30 shows a complete receiver with the relay mounted
on the top near the coherer, together with an inker mechanism for
making a permanent record at the front. The use of the German words
'fritter' and 'klopfer' may be remarked here. Fig. 3.31 shows this receiver as part of a complete transmitting and receiving installation of
about 1907. The receiver pictured in Fig. 3.32 was used by Captain
H.B. Jackson of the British Navy in the years 1896-7.

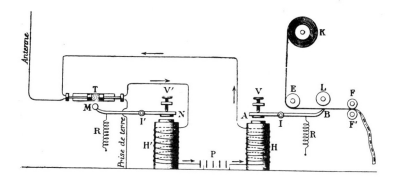

Fig. 3.27 *Branly coherer with tapper (M) and inker (B)*
[Monier, E.: *La Télégraphie sans fil* (Dunod et Pinat, 1913), p. 24]

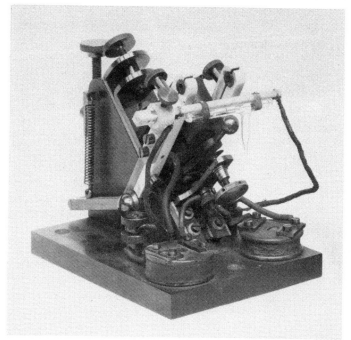

Fig. 3.28 *Marconi's coherer and tapper*
[Science Museum photograph]

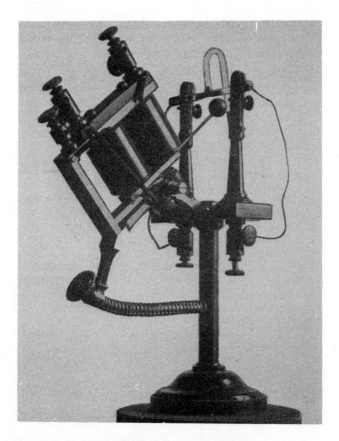

Fig. 3.29   *Blondel's side-pocket coherer with its tapper*
[Collins, *Electrical World and Engineer,* **39**, 1902]

Tapping with a hammer was by no means the only method of decohering. Fleming, for example mounted the coherer on the vibrating reed of the relay itself,[91] while L.B. Miller[92-93] mounted the coherer (B) on a springy bar (marked A in Fig. 3.33). A threaded rod fixed to the bell trembler lay across this bar, and when the bell was ringing, moved to and fro with a sawing motion which provided sufficient vibration to effect the restoration of the coherer. Returning for a moment to Fig. 3.21, Massie's coherer was mounted on a similar springy mounting (3) and a tap from the hammer (8) was sufficient to do the trick in this case.

A somewhat more compact configuration patented by Lodge and Muirhead can be seen in Fig. 3.34.[94-95] The filings lie between two metal plates which are pressed together by the two springs (g'). When

Relais
Fritter
Klopfer
7
48
6
8

Fig. 3.30 *Telefunken receiver with coherer (fritter) and tapper (klopfer) circa 1907*
[Arendt, O.: *Die Electrische Wellentelegraphie* (Vieweg, Braunsch-weig, 1907), facing p.120]

coherence occurs, battery current flows through the lower plate which is made of flexible material, and since this is situated in the field of a permanent magnet mounted below, it experiences a force which jerks the plate momentarily, causing decoherence of the filings.

Another philosophy was to keep the filings in a constant state of mechanical agitation so that the incoming signal had to cohere them against the mechanical movement. This constant agitation was provided by tapping the coherer continuously with a clockwork-driven hammer,[96-97] by rotating the coherer tube about its long axis,[98-101] or even, in one case, by rotating it end over end.[102] When the filings were made from a ferrous metal a rotating permanent magnet could be used to stir them, as in Fig. 3.35, or else an alternating current flowing in a coil wound on the tube could achieve the same effect.[103] Alternatively, a solenoid situated near the tube could simply be used in place of a tapper.[104-107] The battery current would be passed through this solenoid and the magnetic filings would be torn out of coherence by

Fig. 3.31   *Telefunken wireless telegraphy station circa 1907*
[Arendt, O.: *Die Electrische Wellentelegraphie* (Vieweg, Braunsch-
weig, 1907), p.112]

Fig. 3.32   *Coherer receiver used by Captain H.B.Jackson in 1896/7*
[Science Museum photograph]

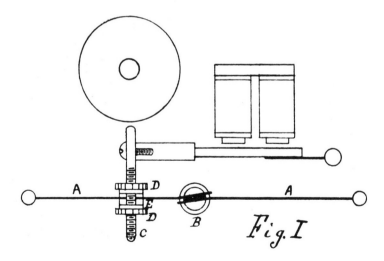

Fig. 3.33   *Decohering arrangement due to Miller*
The bell trembler rubs the threaded rod (C) across the suspension wires
(A) which support the coherer (B)
[British Patent 29 778, 1897]

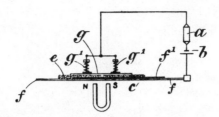

Fig. 3.34   *Lodge-Muirhead decohering arrangement*
Current flowing through flexible plate (*f*) in the magnetic field causes
movement which effects decoherence of the filings (*e*)
[British Patent 18 644, 1897]

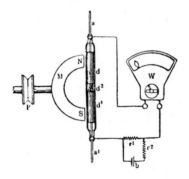

Fig. 3.35   *Agitation of the filings by means of a rotating permanent magnet*
[Mazzotto, D.: *Wireless telegraphy and telephony* (Whittaker, 1906),
p.180]

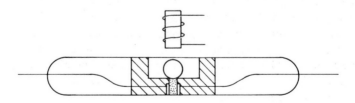

Fig. 3.36   *Shoemaker's coherer with electromagnet and steel ball for decohering*

the magnetic attraction. A variant on this scheme was the tube patented by H. Shoemaker[108] in 1901 (Fig. 3.36). Two L-shaped insulating blocks are included in the tube and the filings rest in the gap between two electrodes built into the bottom limbs. A steel ball normally rests on the two blocks, but when a solenoid is energised via a relay in the coherer circuit the ball is attracted upwards, striking the inside of the tube and giving the necessary mechanical shock. In another version[109] Shoemaker uses a metal ring loosely fitted on the outside of the tube, instead of the internal ball.

Other decohering arrangements are shown in Fig. 3.37–3.39. In the rather primitive arrangement used by Tommasina[110–111] the filings

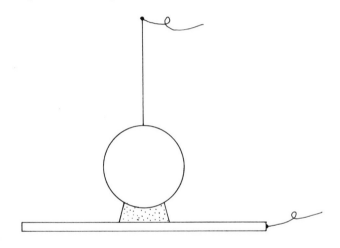

Fig. 3.37 *Tommasina's pendulum filings-coherer.*
The natural sway of the bob ensured decoherence

are arranged between the metal bob of a pendulum and a plate just below it. The filings cohere in the usual way, but it relies on the natural swaying of the pendulum in the room draughts to produce the restoring mechanical movement. In a patent of 1897 Miller[112] specified the use of a rotary coherer, as in Fig. 3.38. A wire (W) was included in the battery circuit. When current flowed (via a relay, if necessary), expansion caused the wire to sag, and it was pulled sideways by a small spring. The cord connecting the wire to the spring also passed over the coherer so that as it sagged it rotated the coherer slightly on its pivots, one of which was actually a contact dipping into the metal filings. This slight twist was sufficient to disturb the coherence and restore it to a condition in which it was ready to receive the next signal. Fig. 3.38*b*, re-

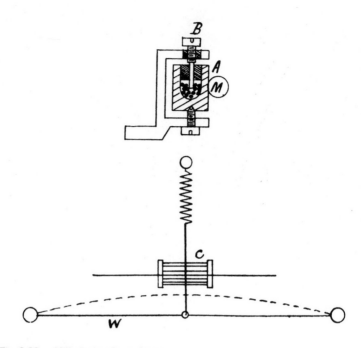

Fig. 3.38   *Miller's rotating coherer*
(a) construction of the coherer
(b) Method of decohering using the sag of a heated wire (W) to rotate the coherer
[British Patent 29 778, 1897]

produced from the patent, shows a horizontally mounted coherer, but the same principle can clearly be applied to the vertical coherer of Fig. 3.38a. Both are described in the patent.

Alfred G. Dell[112-114] published details of the coherer of Fig. 3.39. Two thin pieces of wood were bent to form a funnel, and this was filled with the metal filings. The motion of the filings running out of the bottom past two stiff brass wires (a) inserted into the neck of the funnel maintained a constant state of high resistance between those wires. An oscillatory received voltage applied between the wires caused the filings to cohere against the constant motion, but as soon as the signal ceased, gravity once more caused the particles to run out and destroyed the low resistance. Not surprisingly, he reported that he had great difficulty with the filings tending to clog up in the hole, but he asserted that the use of a smooth material such as glass for the funnel instead of wood would cure this defect, and he suggested making one in the form of an hour-glass which could be inverted when all the filings had run to the bottom.

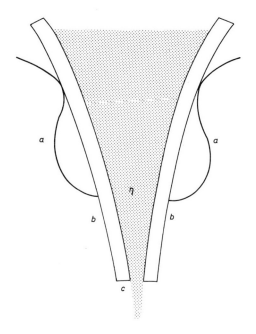

Fig. 3.39   *Dell's 'hour glass' coherer*
[*Electrical World and Engineer,* **33**, 1899, p.839]

Various people claimed that when the particles in the coherer were made from a magnetic substance the provision of a permanent magnetic field along the axis of the tube helped to produce a reliable and sensitive device.[115-119] The very elaborate system used by Tissot is shown in Fig. 3.40, and it illustrates the degree of sophistication and complexity achieved in some of these coherer arrangements.[120] The coherer tube itself is arranged vertically and is free to swing like a pendulum on the knife-edge pivots (*a*). Normally the current from the battery π' flowing through the solenoid B maintains an axial magnetic field, the strength of the field at the actual filings being varied by sliding the coil along the tube until the best position is found. Arrival of a signal in the aerial coheres the tube and completes the circuit of the relay R and battery π causing the relay armature to move from contact δ to contact γ. This causes various other things to happen:

(i)   The magnetising solenoid is switched off, owing to the inter-ruption of its energising current at δ. This is necessary so that the magnetic field, which tends to bind the particles together, will not impede the process of decoherence.

(ii)   The circuit of battery P' and coil A is completed through contact γ.

(iii)   A small iron ring (*b*) mounted on the bottom of the coherer tube is attracted to coil A. The coherer swings from the vertical, and the jerk produced as it hits against a stop returns the coherer to the high-resistance state.

(iv)   As the coherer swings from the vertical the relay energising circuit is broken at contact *a* so that the system resets itself with the magnetising current re-established in the coil B.

(v)   Finally, while the tube is displaced from the vertical the contact β allows battery P to operate the Morse inker or other indicating instrument T.

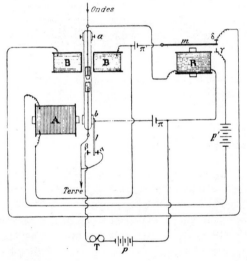

Fig. 3.40   *Tissot's coherer circuit*
[The solenoid (B) is energised when the coherer is in the high-resistance condition and magnetises the *filings* in the tube
[Turpain, A.: *Less applications des ondes electriques* (Carré et Naud, 1902), p.188]

So far in this discussion of decohering methods we have dealt ex-clusively with filings coherers, but the single point coherers also needed to be restored by means of a mechanical shock. Lodge achieved this in his flat-spring coherer of Fig. 3.10 by extending the spring out beyond the clamp (at the left-hand side of the diagram) and allowing a small cog wheel with fine teeth to rotate against it constantly, rather in the manner of a football rattle. The resulting vibrations were just sufficient to maintain the contact in the high-resistance condition except when a cohering voltage was actually present.[121] There was, however, one disadvantage with this arrangement. If a pulse of voltage appeared just at the instant when one of the teeth of the wheel happened to be plucking the spring there was a danger that this pulse would be missed altogether. This was overcome in the later arrangement of Fig. 3.41, where three cohering contacts were operated in parallel.[122] A rotating cam (*d*) flicked each one in turn so that at any instant at least one was ready to receive a signal, provided that the speed of signalling was not too great.

The arrangement which Branly[123–124] used to restore his tripod coherer is shown in Fig. 3.42, a photograph of the actual apparatus being also shown in Fig. 3.43. Coherence of the tripod completed the circuit of relay X, causing solenoid B to attract the armature *d* which struck the coherer mounting, and restored it to the sensitive condition. The strength of the decohering blow could be regulated to some extent by setting the screw *b*. A stylus or pen *p* was fixed to the other side of the armature and made a permanent record of the signal on the moving band of paper. Branly's original paper shows an alternative system where an electromagnet was used to lift the tripod bodily off the plate to effect decoherence.

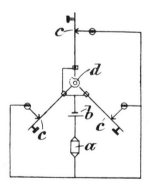

Fig. 3.41 *Rotating cam (d) used to decohere three parallel-connected single-point coherers*
[British Patent 18 644, 1897, Lodge and Muirhead]

Fig. 3.44 shows a single-point coherer attributed to Mareschal, Michel and Dervin, where one of the cohering contacts is the diaphragm of an ordinary telephone earpiece. The circuit connections are not particularly clear in this diagram, but the idea is that the battery current should flow through the coil attracting the diaphragm and breaking the cohering contact. The 'radioscope', a form of coherer patented by S. G. Brown in 1903, achieved decoherence in a similar sort of way.[126-127]

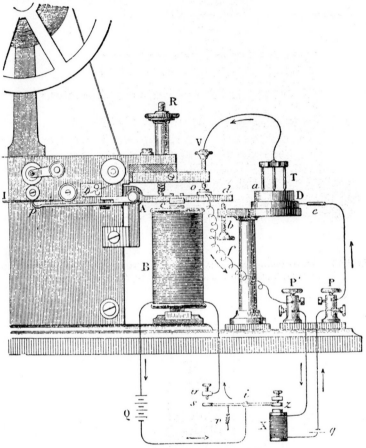

Fig. 3.42   *Branly's receiver using his tripod coherer (T)*
[Monier, E.: *La télégraphie sans fil* (Dunod et Pinat, 7th edn., 1913), p.37]

Self-restoring coherers

In the course of their experiments, various people discovered that imperfect contacts between certain particular substances acted as coherers in the usual way, but they possessed in addition the property of automatic decoherence, or self restoration, as it was commonly called. That is to say, after having been cohered they returned to the high-resistance

Fig. 3.43   *Actual photograph of the receiver of Fig. 3.42.*
[Branly, E., Comptes Rendus, 1902, **134**, p.1198]

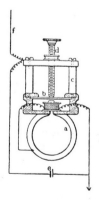

Fig. 3.44   *Single-point coherer attributed variously to Mareschal, Michel and Dervin, having telephone-diaphragm arrangement for decohering.*
[Mazzotto, D.: *Wireless telegraphy and telephony* (Whittaker, 1906), p.181]

condition as soon as the voltage was removed, without any need for mechanical tapping. This was clearly a most desirable property leading to a great simplification in the circuit arrangements. For example, in the course of his extensive investigations Professor Chunder Bose discovered that the metal potassium behaved in this way if it were kept under oil or in a vacuum in order to prevent the formation of an oxide coating.[128]

However, most of these self-restoring coherers, otherwise known as autocoherers, used steel, carbon or mercury in various combinations.[129-134] It is worth noting here that some microphonic coherers used by Hughes in his pre-Hertzian experiments used carbon/carbon or carbon/steel contacts, and he subsequently stated that he had found them to be self restoring. Jervis-Smith,[135-136] Stone,[137] Popov and

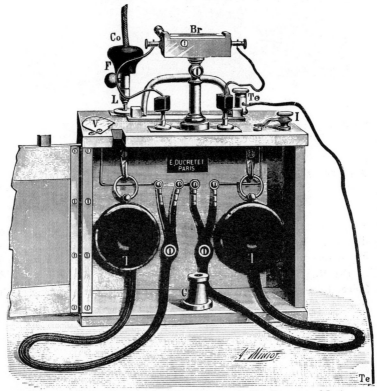

Fig. 3.45   *Version of the Popov receiver constructed by Ducretet of Paris*
The Self-restoring coherer is labelled 'Br'
[Mazzotto, D.: *Wireless telegraphy and telephony* (Whittaker, 1906) p.185]

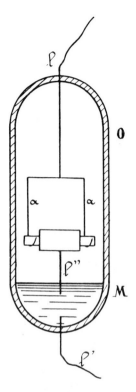

Fig. 3.46   *Minchin's self-restoring coherer using an aluminium/carbon contact*
[Boulanger, J. and Ferrié, G.: *La Télégraphie sans fil et les ondes électriques* (Berger-Levrault, 1907), p.215]

others[138] used carbon granules in place of metal filings and found that they required virtually no mechanical disturbance to decohere them. A Popov receiver of this type constructed by the Ducretet establishment of the Rue Claude-Bernard, Paris may be seen in Fig. 3.45.[139-140] Since there is no tapper, the circuit is very simple, consisting of the coherer in series with a battery and two earpieces. The battery was contained in the box (not visible in this picture); for transporting it about, the coherer (Br) could be mounted on the stand (C) and the front cover closed, making a neat little box with a handle on the top. When in use the coherer was mounted as illustrated and was connected into the circuit by the two plug connectors. The coherer contained either carbon, or, in a second version, steel particles made by crushing tempered ball bearings, and it was mounted on a swivel so that it could be inclined at the best angle for optimum sensitivity. Yet another version used steel needles resting between carbon end plates. One very interesting point to note here is that in this picture of the receiver, which was presumably prepared by Ducretet for advertising purposes, the coherer is indicated by the letters Br (standing for Branly?). Was this a subtle ploy on the part of the French to establish the prior claims of their fellow countryman over their Russian customer?

The effectiveness of this receiver was demonstrated sometime around 1900 when, towards the end of the year a Russian warship broke down near Hohland island in the Gulf of Finland. She could not be extricated before the freeze set in and she was forced to spend the winter there in that very isolated spot. Popov was given the task of establishing wireless communication and he set up three stations, one on the ship, another on the mainland near the town of Kotka, and the third on the ice-breaking vessel *Ermack*. Between the end of January when the installation was completed and April when the vessel was released from the ice a total of 440 messages were exchanged, the installation proving to be very reliable. Indeed it saved the lives of twenty-seven fishermen who had become trapped on a drifting ice-floe by enabling a message to be sent to the *Ermack* which was able to pick them up in time.

Another self-restoring coherer was demonstrated by Minchin in 1904,[141-142] and his device, which used carbon/aluminium contacts, is shown in Fig. 3.46. A small carbon rod rests on two hooks fashioned out of aluminium wire, and contact is made to the rod by means of a platinum wire dipping down into a pool of mercury lying in the bottom of the tube. Blake reported the curious fact that when simply used with a telephone it was sufficiently self restoring to reproduce signals in the telephone without tapping; when used in conjunction with a relay, however, tapping was found to be necessary. Without performing further experiments on such an arrangement it is impossible to explain why this should have been so, but it certainly gives the impression that it was not a particularly reliable affair.

Several well engineered and quite elaborate versions of the carbon/ steel coherer were produced, one of the Ducretet – Popov devices shown in Fig. 3.47 being an example.[143] Here needles made of steel rest on supports (E and E′) made of carbon or steel. The chambr (De) at the top contains a dehydrating agent to keep the air inside dry. In the Shoemaker and Pickard[144] version (Fig. 3.48) very thin needles (34) are held between two thin carbon discs (21 and 25) supported by the metal discs (20 and 24). The black discs (26 and 27) are insulating spacers to keep the whole device central in the tube, and the space (29) at one end communicating with the interior by means of holes (30) once again contains a substance to ensure dryness of the air. To quote

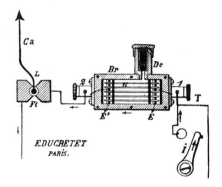

Fig. 3.47 *Popov–Ducretet self-restoring coherer employing steel rods resting on metal and carbon supports*
[Mazzotto, D.: *Wireless telegraphy and telephony* (Whittaker, 1906), p.177]

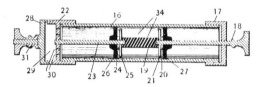

Fig. 3.48 *Shoemaker and Pickard self-restoring coherer using steel needles (34) supported between carbon discs (21 and 25)*
[Sewall, C.H.: *Wireless telegraphy* (Crosby-Lockwood, 1904), p.158]

from their patent specification it formed 'a wave-responsive device of great delicacy and sensitiveness in responding to electrical radiation and has the desirable property of regaining its normal condition after the cessation of the influence of the electric waves'.

A very neat version of the carbon-granule coherer developed by Tommasina[145–146] is shown in Fig. 3.49. The coherer itself is so small as to be mounted inside a normal telephone earpiece. The only external connections needed were to aerial and earth and to a battery. In 1899 Tommasina also constructed a coherer which consisted of a drop of mercury held between two brass carbon, or iron electrodes, but curiously enough he does not seem to have mentioned any self-restoring properties.[147–148] Perhaps the best known of all the mercury coherers was that devised by P.Castelli and/or the Marquis Luigi Solari,[149–153] consisting of drops of mercury contained between iron or carbon plugs in various combinations, as in Fig. 3.50. A great deal of bitter correspondence may be read in the journals of the period concerning which of the two persons just named actually was the inventor of this coherer.

The world at large seems to have evaded this question by referring to it simply as the 'Italian Navy' coherer since both were officers in that service. Marconi used this type of coherer[154] (among others) in his famous trans-Atlantic trials of 1901, but seems to have been unimpressed by its general performance, for in a lecture to the Royal Institution in June 1902 he said[155-156]

> These non-tapped coherers have not been found to be sufficiently reliable for regular or commercial work. They have a way of cohering permanently when subjected to the action of strong electrical waves or atmospheric electrical disturbances, and have also an unpleasant tendency towards suspending action in the middle of a message. The fact that their electrical resistance is so low and always varying when in a sensitive state causes them to be unsatisfactory with my system of syntonic wireless telegraphy. These coherers are, however, useful if employed for temporary tests in which complete accuracy of message is not all-important.

R.P.Howgrave-Graham[157-158] also seems to have suffered, for he mentions that when the Italian Navy coherer was in series with a telephone earpiece, noises were heard 'like oil frying in a distant frying-pan'.

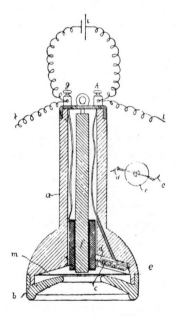

Fig. 3.49   *Tommasina carbon-granule coherer (C) mounted inside a telephone earpiece*
[Mazzotto, D.: *Wireless telegraphy and telephony* (Whittaker, 1906), p.183]

Another mercury-based coherer for which excellent performance was claimed was the 'Tantalum' detector of L.H. Walter[159–160] (Fig. 3.51*a*). Some mercury (M) is contained in a sealed glass envelope, and a platinum wire (P) sealed into the glass makes contact with it. Another platinum wire (P_1) sealed through the glass has attached to it a very fine piece of tantalum wire the end of which just touches the surface of the mercury. This contact was found to exhibit the property of coherence, and in addition, was self-restoring. According to Walter's figures this had a ratio of high resistance to cohered resistance as great as 30 : 1, compared with a figure quoted for the Italian Navy coherer of only 3 : 1. One senses, perhaps, the possibility of a slight exaggeration here, but even so it seems to have made a very effective detector. On the other hand, it was certainly not a very rugged device and would have been useless on board ship where the mercury would have been swilling around the bowl. For this reason he devised the portable version shown in Fig. 3.51*b*. Here a very small drop of mercury is completely contained in an insulating enclosure (G, I and I_1). The ordinary ohmic contact is established by means of a piece of platinum wire sticking up into the mercury, but the cohering contact is made by embedding the tantalum wire in a piece of glass (B). The end of this glass is then ground flat, exposing a flat disc of tantalum which is brought into contact with the mercury. One is naturally led to ask why the comparatively obscure metal tantalum was used; the answer is, quite simply, that he tried many other metals but claimed that with the possible exceptions of molybdenum and zirconium, tantalum was the only one which gave such excellent performance.

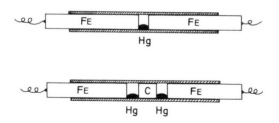

Fig. 3.50 *Two versions of the Italian Navy coherer using mercury between iron or carbon plugs*

Before leaving the subject of coherers we must return for a moment, as promised, to say something about *how* they work; i.e. about the physical effects on which they depend. At the time when they were widely used (broadly speaking, around the turn of the century) there were several theories to explain how the sudden drop in resistance came about. Some were rather vague and spoke of such things as 'molecular rearrangement',[161] 'the electric discharge filling up the intermolecular spaces'[162] or 'condensing air'.[163] Setting aside such intangible ideas, there seemed to be three main theories. The first assumed that normally

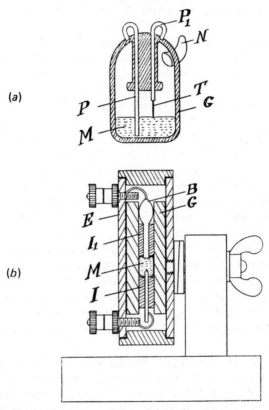

(a)

(b)

Fig. 3.51 *Walter's tantalum detector using a tantalum (T)/mercury (M) contact*
(*a*) Original version, (*b*) Portable form
[*Proc. Roy. Soc.*, **81A**, 1908, pp.3 and 6]

the filings or metal electrodes lie in loose contact with one another, but when a voltage is applied across them they experience electrostatic forces which cause them to come together and form chains.[164-173] Each particle is now pressed into close contact with its neighbours thus establishing a good metal-to-metal contact. Some writers claimed to have seen movements of this sort when the filings were examined under a microscope, and claimed also to have seen sparks jumping from particle to particle[174] and the formation of chains glowing red-hot within the mass of filings.[175] Others were very sceptical, arguing that it was impossible to conceive of the production of the large forces necessary to move filings around under the influence of the very small voltages to which the coherer was normally subjected.[176] Most coherers were, after all, able to operate quite successfully with voltages of only a few tenths of a volt.

It was generally realised that the insulating film of oxide which formed on the surface of the contacts or the filings was of paramount importance in producing a satisfactory device.[177-179] One inventor even proposed situating his coherer, which consisted of a contact between copper and lead, over a flame so as to maintain this all-important film.[180-181] Unoxidised particles were found to make poor coherers, yet on the other hand very heavy oxidation also made for insensitivity and sluggish action. It was this fact that led Marconi and others to evacuate their coherers of air in order to prevent continuing oxidation which gradually reduced their effectiveness. Others immersed their filings in oil for the same reason.[182] It is also interesting to note that satisfactory coherers were made with particles which had sulphide coatings[183] or were even covered with wax or resin films to provide a thin insulating layer.[184-186] As previously mentioned, the noble metals which tarnished slowly were generally found to be unreliable in action. Recognising the extreme importance of this surface layer, and bearing in mind that fact that the Sanskrit word for skin is 'twach', Professor Chunder Bose[187] suggested that the coherence phenomenon should be called the 'electric touch', a delightful idea which, alas, never caught on.

The second theory to explain coherence was that the mechanical forces caused rupture of the very thin oxide film on the particles followed by welding together at the minute contact points thus created.[188-189] A variant on this theme was that a tiny hole was formed in the insulating layer, followed by evaporation of a tiny amount of metal which coated the inside of the hole forming a conducting bridge across the insulation.[190-192] The two theories mentioned so far are not incompatible in that breakdown of the insulation could occur after the formation of particles into chains.

The third theory was that the resistance drop was purely[193-195] thermal in origin. Most metals have a positive coefficient of change of resistance with temperature — i.e. when you heat them their resistance rises. It was noticed that many of the metals which were satisfactory for use in coherers had oxides which possessed negative resistance coefficients. It was thought that localised heating at the points of contact led to a decrease in the resistance of the oxide film at those points. In certain cases this could lead to thermal runaway where decreasing resistance allowed more current to flow, which in turn decreased the resistance still further and so on, ending with a large flow of current which could weld the particles together.

Argument raged at the time as to which of these theories was correct, but other detectors came along and coherers fell into disuse before the matter could be finally resolved. There have been some modern investigations into the subject.[196] Suffice it to say here that depending on the conditions in any particular case, and especially on the thickness of the oxide coating, all three mechanisms can play a part as the present author's own experiments have shown.[197]

To summarise: the coherer in all its various forms was a useful and sensitive detector which was widely used for wireless telegraphy until better methods of detection displaced it. The last word on the subject must surely be left to W.H. Eccles[198] who, in a spirited defence of the coherer in 1907, explained 'the coherer's reputation for erratic behaviour is wholly due to the performance of samples made in a slipshod manner. Coherers so made might be aptly compared with homemade false teeth'.

References

1   BRANLY, E.: *Electrician,* **27**, 1891, p.221
2   VON LANG, V.: *Wied. Ann.,* **57**, Pt. I, 1895 – 6, p.34
3   MUNK AF ROSENSCHÖLD, P.S.: *Poggendorf's Ann. d Phys.,* **34**, Pt.3, 1835, p.437
4   MUNK AF ROSENSCHÖLD, P.S.: *Poggendorf's Ann. d Phys.* Pt.3, 1838, p.193
5   GUTHE, K.E.: *Electrician,* **54**, 1904, p.92
6   EKELÖF, S.: *Marconi centenary convention.* Acad. Nazionale dei Lincei, Roma, 1974, p.184
7   LODGE, O.J. *et al.: Electrician,* **40**, 1897, p.86
8   CALZECCHI-ONESTI, T.: *Nuovo Cimento,* Ser. 3, **17**, 1885, p.38
9   *J. Soc. Telegraph. Eng.,* **16**, 1887, p.156
10   OLIVETTI, C.: *Electrical World and Engineer,* **34**, No.23, 2nd Dec. 1889
11   BRANLY, E.: *Lumière Électrique,* **40**, 1891, p.301
12   BRANLY, E.: *Éclairage Électrique,* **3**, Ser. 2, 1895, p.361
13   BRANLY, E.: *Éclairage Électrique,* **3**, Ser. 2, 1895, p.230
14   *J. Soc. Telegraph Eng.,* **24**, 1895, p.639
15   BRANLY, E.: *Electrician,* **27**, 1891, p.221 and p.448
16   *Electrician,* **45**, 1900, p.878
17   BLONDEL, A., and FERRIÉ, G.: *Electrician,* **46**, 1900, p.21
18   HOLM, R.: *Electric contacts handbook* (Springer Verlag 3rd Edn., 1958), p.403

19 Reference 5
20 LODGE, O.J.: *Signalling through space without wires* (Electrician, 3rd edn. reprint of Royal Institution lecture of 1894), p.22
21 LODGE, O.J. and MUIRHEAD, A.: British Patent 18 644, 1897
22 ECCLES, W.H.: *Electrician*, **47**, 1901, p.682
23 HALE, H.: *Electrical World and Engineer*, **41**, 1903, p.823
24 BRANLY, E.: *Comptes Rendus*, **134**, 1902, p.1197
25 *Science Abstracts*, **5**, 1902, no.2092, p.852
26 *Science Abstracts* (B), 7, 1904, No.379, p.155f
27 MONIER, E.: *La Télégraphie sans fil* (Dunod et Pinat, 7th Edn., 1913), p.91
28 OBENBACH, F.L.: *Western Electrician*, **29**, 1901, p.349
29 *Science Abstracts*, **5**, 1902, no. 718, p.309
30 *Science Abstracts*, **3**, 1900, no. 2034, p.802
31 *Electrical Magazine*, **1**, 1904, p.624
32 JENTSCH, O.: *Telegraphie und Telephonie ohne draht* (Springer, 1904), p.176
33 MASKELEYNE, J.N.: British Patent 16 113, 1903
34 ECCLES, W.H.: *Electrical Engineering*, **1**, 1907, p.241
35 *Revue Electrique*, **3**, 1905, p.184
36 *Electrical World and Engineer*, **34**, 1899, p.317
37 Reference 20
38 BLAKE, G.G.: *History of radio telegraphy and telephony* (Chapman and Hall, 1928), p.62
39 Reference 38, p.64
40 Reference 20, p.27
41 POPOV, A.: *Electrician*, **40**, 1897, p.235
42 *Science Abstracts*, **2**, 1899, no. 1164, p.520
43 BRANLY, E.: *Comptes Rendus*, **128**, 1899, p.1089
44 *Electrician*, **43**, 1899, p.111
45 ORLING, A. and BRAUNERHJELM, C.G.G.: British Patent 1 866, 1899
46 ORLING, A. and BRAUNERHJELM, C.G.G.: British Patent 1 867, 1899
47 TURPAIN, A.: *Les applications pratriques des ondes électriques* (Carré et Naud, 1902), p.256
48 Reference 32, p.177
49 BOSE, J.C.: *Phil. Mag.*, **43**, 1898, p.55
50 BOSE, J.C. *Electrician*, **36**, 1895, p.291
51 BOTTONE, S.R.: *Wireless telegraphy and Hertzian waves* (Whittaker, 4th edn., 1910), p.52
52 TAYLOR, A.H.: *Physical Review* (New York), **16**, 1903, p.199
53 *Electrical Engineer*, (London), **31**, 1903, p.581
54 ROBINSON, P.E.: *Annalen der Physik* (Leipsig), **11**, no. 2, 1903, p.755
55 BRANLY, E.: *omptes Rendus*, **127**, 1898, p.1206
56 *Science Abstracts*, **2**, 1899, no. 693, p.294
57 DORN, E.: *Wied. Annalen*, **66**, 1898, p.146
58 *Science Abstracts*, **2**, 1899, no. 106, p.38
59 Reference 47, p.126
60 *Electrician*, **43**, 1899, p.111
61 Reference 38, p.70
62 BRANLY, E.: *Electrician*, **27**, 1891, p.448

63   BOSE, J.C.: *Proc. Roy. Soc.* **(London), 65**, 1899, p.166
64   Reference 20
65   BLONDEL, A.E.: *Electrician,* **43**, 1899, p.277
66   POPOV, A.S.: *Electrician,* **40**, 1897, p.235
67   MASSIE, W.W.: US Patent 800 119, 1905
68   *Revue Electrique,* **4**, 1905, p.234
69   MARCONI, G.: *JIEE,* **28**, 1899, p.273
70   Reference 47, p.143
71   MARCONI, D.: *My father Marconi* (Muller, London, 1962), p.60
72   References 65 and 47
73   COLLINS, A.F.: *Electrical and Wireless Engineer* (New York), **39**, 1902, p.956
74   *Electrical World and Engineer,* **34**, 1899, p.59
75   ADAM, M.: *Cours de T.S.F.* (Editions Radio-Home, undated), p.116
76   MAZZOTTO, D.: *Wireless telegraphy and telephony* (Whittaker, 1906), p.174
77   Reference 76, p.173
78   BOULANGER, J. and FERRIÉ, G.: *La Télégraphie sans fil* (Berger-Levrault, 1907), p.83
79   GAVEY, J.: *JIEE,* **36**, 1905, p.15
80   MINCHIN, G.: *Phil. Mag.,* **37**, Ser. 5, 1894, p.90
81   *J. Soc. Telegraph Eng.,* **23**, 1894, p.251
82   Reference 51, p.106
83   *Electrician,* **52**, 1903, p.42
84   Reference 38, p.65
85   POPOV, A.: *Electrician,* **40**, 1897, p.235
86   MARCONI, G.: British Patent 12 039, 1896
87   Reference 20, p.62
88   Reference 47, p.366
89   ECCLES, W.H.: *Wireless* (Thornton Butterworth, 1933), p.55
90   Reference 27, p.24
91   FLEMING, J.A.: *Principles of electric wave telegraphy and telephony* (Longmans, 3rd edn. 1916), p.484
92   MILLER, L.B.: British Patent 29 778, 1897
93   Reference 51, p.43
94   LODGE, O.J. and MUIRHEAD, A.: British Patent 18 644, 1897
95   Reference 76 p.180
96   *Electrical Magazine,* **2**, 1904, p.166
97   HOWGRAVE-GRAHAM, R.P.: *Wireless telegraphy for amateurs* (Percival Marshall, London, 1907[?]) p.153
98   *Electrical Review,* (London), **42**, 1898, p.535
99   RUPP, H.: *Electrical Engineer* (New York), **25**, 1897, p.543
100 Reference 38, p.63
101 ECCLES, W.H.: *Handbook of wireless telegraphy and telephony* (Benn, 2nd Edn., 1918), p.283
102 SEWALL, C.H.: *Wireless telegraphy* (Crosby Lockwood, 1904), p.152
103 Reference 76, p.180
104 SHOEMAKER, H.: British Patent 12 816, 1901
105 TOMMASINA, T.: *Comptes Rendus,* **128**, 1899, p.1225
106 *Electrical World and Engineer,* **34**, 1899, p.389

107 Reference 17
108 SHOEMAKER, H.: British Patent 11 214, 1901
109 SHOEMAKER, H.: British Patent 14 919, 1901
110 TOMMASINA, T.: *Comptes Rendus,* **127,** 1898, p.1014
111 *Science Abstracts,* **2,** 1899, no. 692, p.293
112 MILLER, L.B.: British Patent 29 778, 1897
113 DELL, A.G.: *Electrical World and Engineer,* **33,** 1899, p.839
114 *Science Abstracts,* **2,** 1899, no. 1719, p.758
115 *Scientific American,* **83,** 1900, p.87
116 TISSOT, C.: *Comptes Rendus,* **130,** 1900, p.902
117 Reference 17
118 Reference 76, p.179
119 TORIKATA, W.: *Electrician,* **65,** 1910, p.940
120 Reference 47, p.188
121 LODGE, O.J.: British Patent 11 575, 1897
122 Reference 21
123 BRANLY, E.: *Comptes Rendus,* **134,** 1902, p.1197
124 Reference 27, p.37
125 Reference 76, p.181
126 BROWN, S.G.: British Patent 750, 1902
127 *Electrician,* **52,** 1903, p.85
128 BOSE, J.C.: *Proc. Roy. Soc.* (London), **65,** 1899, p.116
129 BOSE, J.C.: *Electrician,* **43,** 1899, p.441
130 *Electrical World and Engineer,* **39,** 1902, p.1017
131 Resference 69 (Discussion)
132 TOMMASINA, T.: *Comptes Rendus,* **128,** 1899, p.666
133 SHOEMAKER, H.: British Patent 23 574, 1903
134 VON LANG, V.: *J. Soc. Teleg Eng.,* **25,** 1896, p.255
135 Reference 102, p.157
136 JERVIS-SMITH, F.J.: *Electrician,* **40,** 1897, p.84
137 STANLEY, R.: *'Textbook on wireless telegraphy'* (Longmans, 2nd edn., 1919), p.300
138 BROWN, A.C.: *Electrician,* **40,** 1897, p.116
139 Reference 76, p.185
140 Reference 51, p.60
141 Reference 78, p.215
142 Reference 38, p.69
143 Reference 76, p.177
144 Reference 102, p.158
145 Reference 76, p.183
146 TOMMASINA, T.: *Electrical World and Engineer,* **35,** 1900, p.835
147 TOMMASINA, T.: *Electrician,* **43,** 1899, p.111
148 Reference 38, p. 71
149 *Electrician,* **49,** 1902, p.387
150 GUARINI, E.: *Electrician,* **51,** 1903, p.360
151 SOLARI, L.: *Electrician,* **51,** 1903, p.502
152 BANTI, A.: *Electrical Engineer,* **30,** 1902, p.55
153 *Electrical Review,* **51,** 1902, p.968
154 MARCONI, G.: British Patent 18 105, 1901
155 MARCONI, G.: Lecture to Royal Institution, 13th June 1902, reprinted in

Wireless Telegraphy (Royal Institution Library of Science, Applied
Science Publishers, 1974), p.81

156   Reference 102, p.161
157   Reference 97, p.156
158   Reference 153
159   WALTER, L.H. *Proc. Roy. Soc.* (London), **81 A**, 1908, p.1
160   Reference 91, p.486
161   TURNER, D.: *Electrician,* **29,** 1892, p.432
162   BRANLY, E.: *Comptes Rendus,* **122,** 1896, p.230
163   SHAW, P.E.: *Phil. Mag.,* Ser 6, **1**, 1901, p.265
164   ERSKINE-MURRAY, J.: *JIEE,* **36,** 1906, p.384
165   LODGE, O.J.: *Phil. Mag.,* **37,** 1894, p.94
166   *J. Soc. Telegraph Eng.,* **23,** 1894, p.252
167   TOMMASINA, T.: *Comptes Rendus,* **129,** 1899, p.40
168   ECCLES, W.H. *Electrician,* **47,** 1901, p.682
169   HANCHETT, G.T.: *Electrical Review* (New York), **42,** 1903, p.599
170   Reference 163
171   SUNDORPH, T.: *Wied. Annalen,* **68,** 1899, p.594
172   LODGE, O.J.: *Phil. Mag.,* **37,** 1894, p.94
173   *Electrical World and Engineer,* **34**, 1899, p.317   1899, p.317
174   MALAGOLI, R.: *Nuovo Cimento,* **8,** 1898, p.109
175   TOMMASINA, T.: *Comptes Rendus,* **127,** 1898, p.1014
176   BRANLY, E.: *Electrician,* **27,** 1891, p.448
177   BROWN, A.C.: *Electrician,* **40,** 1897, p.166
178   GUTHE, K.E., and TROWBRIDGE, A.: *Phys. Rev.,* **11,** 1900, p.22
179   GODDARD, R.H.: *Phys. Rev.,* **34,** 1912, p.423
180   HORNEMANN, M.: *Elektrotech Zeit.* **25**, 1904, p.861
181   *Revue Electrique,* **2**, 1904, p.278   *ue,* **2,** 1904, p.278
182   PICCHI, A.: *Revue Electrique,* **3,** 1905, p.88
183   *Science Abstracts,* **2**, 1899, no.457, p.179', no.457, p.179
184   Reference 80
185   Reference 176
186   *Science Abstracts,* **2**, 1899, no.965, p.428
187   BOSE, J.C.: *Proc. Roy. Soc.,* **66,** 1899/1900, p.450
188   Reference 17
189   LODGE, O.J.: *Electrician,* **40,** 1897, p.90
190   Reference 176
191   VON GULIK, D.: *Wied. Ann.,* **66,** 1898, p.136
192   HAERDEN, T.: *Electrical World and Engineer,* **35,** 1900, p.672
193   ECCLES, W.H.: *Electrician,* **71,** 1913, pp.901 and 1058
194   ECCLES, W.H.: *Electrician,* **65,** 1910, p.727
195   *Electrical Engineering,* **6**, 1910, pp.410 and 523
196   Reference 18
197   PHILLIPS, V.J.: IEE Conference on history of electrical engineering,
         Manchester, 1975, paper 13
198   ECCLES, W.H., *Electrical Engineering,* **1,** 1907, p.241

Chapter 4

Electrolytic detectors

As the name implies, the detectors to be described in this section depended on effects which occur in electrolytes — i.e. in conducting solutions. Broadly speaking, apart from a couple of oddities which will be dealt with at the end, they fall into two categories. In the first category are devices which normally possess a low resistance which increases under the action of high-frequency oscillations. As explained in the previous Chapter, these were generally referred to as 'anticoherers' since the resistance changes in the opposite sense to that in coherers, (although J.A. Fleming regarded this word as being 'uncouth').[1] Devices in the second category have resistances which decrease on application of the oscillations, and contemporary writers quite often regarded them simply as being a particular sort of coherer.

The earliest of the anticoherers seems to have been that known as Schäfer's plate. A thin film of silver was deposited on a piece of glass and was then scratched across with a thin diamond point, producing two sections separated by a very fine gap.[2] Schäfer and another experimenter named Marx[3-4], who subsequently looked into the matter further, found the rather surprising result that the resistance between the two sections was not infinitely great, as might have been expected. In some cases it was as low as 40 Ω. Investigation with a microscope revealed that there were minute scraps of silver produced during the scribing process left lying in the gap, and that when a potential difference of about four volts was applied between the sections these jumped backwards and forwards conveying electric charge from side to side and allowing a small current of flow. This phenomenon is sometimes described in older text-books as the 'electric hail' effect.[5] An additional feature was that these tiny particles tended to link together to form chains or trees across the gap. The completion of such a chain formed a metal-to-metal connection between the sections and a very low resistance resulted.

An even more curious thing was that if the two sections were connected between aerial and earth and occurrence of a high-frequency voltage oscillation had the effect of arresting this movement of particles and disrupting the chains so that the resistance increased markedly. A telephone receiver in the circuit of the local battery rendered this cessation of current audible, and such an arrangement was said to have been used in Germany to receive wireless telegraphy signals over a distance of 95 km.

It might be thought odd that we have started a chapter on electrolytic devices by describing a device which is completely dry; thereby hangs a tale. In 1898 Neugschwender performed similar experiments and found that the operation was greatly improved by breathing upon the gap and creating a film of moisture between the sections.[6] Indeed he went so far as to insist that moisture was necessary before anticoherer action would occur at all, in spite of Marx's claim that he had taken precautions to ensure that his plate *was* completely dry. Some letters setting out the arguments were exchanged on the subject. These were written in German but may be followed in English translation in *The Electrician* of 1901.[7-10] The truth of the matter seems likely to have been that both parties in the dispute were probably correct. The dry plate relied on the small scraps of silver produced in the process of cutting the slit, whereas the wet plate used particles of silver resulting from electrodeposition from one section to the other. The whole phenomenon in the wet plate was beautifully, almost lyrically, described by De Forest[11] (who, incidentally, found that a coating of tin gave better results than silver):

> With the telephone to the ear and the eye to the microscope, the action, thus doubly observed, affords in fact one of the most fascinating, most beautiful pastimes (as I may well term it) ever granted to the investigator in these fields. When the local E.M.F. is first applied to the gap, minute metallic particles, all but invisible even with a thousand power lens, are seen to be torn off from the anode, under the stress of the electric forces, apparently mechanical in action; and these dust-like particles, floating in the fluid, move across to the cathode, some rapidly, some slowly, by strange and grotesque pathways, or directly to their goal. Tiny ferry-boats, each laden with its little electric charge, and unloading its invisible cargo at the opposite electrode, retract their journeyings, or caught by a cohesive force, build up little bridges, or trees with branches of quaint and crystalline patterns. During

this formative period (lasting perhaps for half a minute) the ear hears an irregular boiling sound, and the average deflection of the galvanometer indicates a gradual decrease of resistance, until one or more of these tin trees or tentacles has been built completely across the gap. The silence ensues until current across the bridge is suddenly increased, as by the Hertzian oscillation from an electric spark made in the neighbourhood, or even from a source of so low frequency as the ordinary 60-cycle alternating current. Instantly, all is commotion and change among the tentacles, and especially where these join the cathode. Tiny bubbles of hydrogen gas appear, and, enlarging suddenly, break or burst apart the bridges, while the click in the telephone indicates the rupture of the current's path. Yet they are persevering, these little pontoon ferry-men, and instantly reform, locking hands and hastening from their sudden rout back to build new paths and chains. So the process continues, the local current re-establishing, the electrical oscillations breaking up its highways of passage, with furious bubblings and agitations — a vertiable tempest in a microscopic teapot. The hydrogen gas having, of course, twice the volume of the oxygen, is most in evidence and therefore the rupture of the tentacles occurs chiefly at the cathodic terminal, and where segregated branches of the tin trees are broken off the bubbles are generally noticeable at the cathode. The oxygen, to a large extent, enters into chemical combination with the tin, and after the slit has been used for some time greyish deposit of stannous oxide may be scraped from the anode.

De Forest goes on to say that

one fact must be borne in mind, that the fine tentacles (whose diameter, by the way, is of the order some hundred-thousandths of an inch) do not come into actual metallic contact with the anode terminal. A film of electrolyte of almost molecular thickness must exist between the two conducting normally by electrolytic ionisation and conduction, yet easily decomposed and transformed by a sudden increase in current into an insulating gaseous film, the expansion of which still further increases the resistance of the gap.

This, of course, was only a theory since it is not possible to observe a molecular thickness with an ordinary microscope. It is also possible to assume that when the chain is almost completed across the gap only a tiny extra whisker of metal is necessary to complete the bridge, an amount so small as to be invisible even under a microscope. The effect of an oscillatory voltage is then to rupture this fine connection. It is not possible now to be certain which is the true explanation; further experiments would have to be performed to settle the matter. The interested reader should also refer to a paper by the present author in the Proceedings IEE Conference on the History of Electrical Engineering,

Nottingham, 1978.

As R.P. Howgrove-Graham[12] remarked in his eminently practical way, 'As the moisture dries rather rapidly and it is tiresome to have to breathe on it at intervals, a piece of wet cotton wool may be placed near to, but not in contact with, the gap'. To avoid this particular problem other users coated the plate with a layer of celluloid or collodion, taking care not to fill up the gap completely but to leave cavities in which the action, dry or moist, could take place.[13]

Tommasina used a rather similar type of detector. He suspended a metal pendulum bob over a metal disc, both having been previously coated with copper by electrodeposition.[14-15] The whole thing was immersed in distilled water. On passing a current between them a black deposit of cuprous oxide was formed and chains of particles grew and bridged the gap. If the distance between the bob and the disc was made very small, he claimed that the oscillatory voltage from an aerial circuit was sufficient by itself to cause the formation of chains, so that in this case the resistance decreased and the apparatus acted as a coherer, not

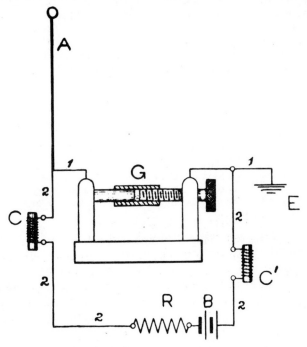

Fig. 4.1   *De Forest's 'responder'*
The active paste was contained in the gap (G)
[Fleming, J.A.: *Principles of electric wave telegraphy and telephony* (Longmans, 3rd edn., 1916), p.508]

an anticoherer.

The most practical and convenient of the anticoherers was that known as the De Forest (or De Forest–Smythe) responder shown in Fig. 4.1.[16-19] It was constructed rather like a filings coherer, there being two metal plugs contained in an insulating tube with a gap of about 1/100 in between them. This gap was filled with one of several mixtures which the inventor referred to as 'goo'. A typical mixture was

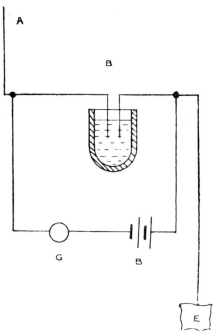

Fig. 4.2   *Pupin's electrolytic detector*
[Erskine-Murray, J.: *Handbook of wireless telegraphy* (Crosby-Lockwood, 4th edn., 1913), p.127]

a paste of litharge (lead oxide), glycerine, water and metal filings. The filings were intended to act as secondary electrodes within the gap. Under the influence of a locally applied direct voltage crystalline lead is produced in such a paste and this builds up into trees or chains across the gap, like the silver particles in Schäfer's plate. When a chain is completed, deposition ceases as the gap is then effectively short circuited. Once again a received pulse of oscillations breaks the chain causing a click to be heard in a telephone receiver which may be connected in the battery circuit. The coils C and C' of Fig. 4.1 perform the same function as those in the coherer circuits of Chapter 3, namely to prevent the shorting of the oscillations by the battery circuit.

The second group of electrolytic detectors made use of polarisation effects in the electrolytes. An early version (about 1899) used by Pupin is shown in Fig. 4.2.[20-22] Two platinum electrodes are immersed in a solution of dilute nitric or sulphuric acid, and a small e.m.f. from a battery is applied between them. Electrolysis then takes place, hydrogen being evolved at the cathode (-ve), and oxygen at the anode (+ve). After a short while polarisation occurs. Gas collects at the electrodes making the passage of current more difficult. If the applied e.m.f. is sufficiently large the polarisation can be overcome and gas production continues, but the optimum condition in this type of detector was when the e.m.f. was just insufficient to overcome the polarisation. A high-frequency oscillation applied between the electrodes seemed to have the property of reducing the polarisation so that conduction would begin again, the increase of current being heard in a telephone or observed on a galvanometer in the usual way. As soon as the oscillations ceased, the gas film formed again and the current fell almost to zero. Pupin himself and other experimenters found that the action was much more marked if one of the electrodes was made much smaller than the other. Some seemed to think that it did not matter much which was the smaller electrode, but on the whole the consensus was that it was better to make the smaller electrode positive as shown in Fig. 4.3. The large electrode was frequently made of lead in order to resist attack by the acid. Several ways of producing a very small diameter electrode were tried. These generally made use of the so-called Wollaston[23-25] wire produced by the method invented by W.H. Wollaston in 1827. In this process a thin piece of platinum wire is first coated with silver to give it strength, and the composite wire is then drawn down through a series of dies until it is as fine as possible. The tip of the wire is then dipped in strong nitric acid or exposed to acid

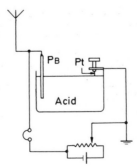

Fig. 4.3 *Fine-point electrolytic detector*

fumes which will dissolve away the silver leaving an extremely fine piece of platinum. In this way diameters as small as 1/100 mm may be obtained with a little care.

A cell made in this way formed a very sensitive detector, a type very popular between the years 1906–13. W.H. Eccles[26] stated that 'the sensitiveness of the detector for a single dot which was just audible in a good pair of telephones has been given as 0·001 erg. A good instrument with suitable telephones and average ears will yield unit audibility [dots and dashes just distinguishable] under the stimulus of about 6 x 10^{-10} watt'. Fig. 4.4 shows a properly engineered version of the detector having screw-thread adjustments to facilitate precise manipulation of the depth of immersion of the point for optimum results.[27]

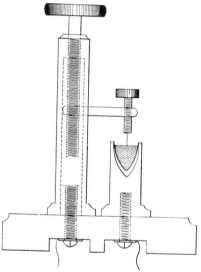

Fig. 4.4   *Fully engineered version of the fine-point electrolytic detector having screw thread adjustment of the depth of immersion of the fine point* [Collins, A.F.: *Manual of wireless telegraphy* (Wiley, Chapman & Hall, 1906), p.52]

The version of Fig. 4.5 was designed to be built by the home constructor, and made use of such homely bits and pieces as shellac and a cork or a rubber bung.[28] The cork was intended to be inserted into a bottle which would act as a stand and keep the whole thing upright. A commercial version due to Ducretet and Roger is shown in Fig. 4.6*a*.

Although a very sensitive detector, in this form it was clearly not suitable for use in any but the most stable environment since any slight movement of the electrolyte would have destroyed the very fine point-to-acid contact at once. Various attempts were made to improve it by embedding the Wollaston wire in a glass rod and grinding the end so that only a minute disc of platinum was exposed — rather like the port-

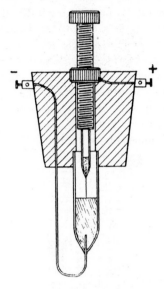

Fig. 4.5   *Electrolytic detector for the home constructor*
[de Blonay, R.: *Wireless World,* **2**, 1915, p.762]

able version of Walter's tantalum detector described in Chapter 3.[29–31] This rod could then be dipped well into the electrolyte to accommodate the effects of surface motion. Another idea was to wax a fine piece of platinum wire leaving only the tiniest tip of the metal exposed. However, neither of these two methods seems to have been very successful, being found to result in diminished sensitivity.[32]

Apart from the effects of motion, evaporation or leakage of the acid was a problem which led to the need for continual readjustment of the point, and, in addition, the fine point itself tended to be very susceptible to damage by strong atmospherics. This was very inconvenient for the operator, and to overcome such frustrations the German company Telefunken gesellchaft für Drahtlose Telegraphie introduced the form shown in Fig. 4.7.[33] Three detectors were mounted together, 120° apart on a common central tube. The glass envelopes of all three were internally connected, but there was only sufficient acid for one cell. In the event of failure the whole assembly was rotated to bring a fresh cell to the bottom, so that normal service could be resumed with the minimum of delay.

Fig. 4.6 *(a)* Electrolytic detector made by Ducretet and Roger of Paris
(b) De Forest electrolytic receiver
[Science Museum photographs]

Several people seem to have been connected with the development of these electrolytic detectors and in the literature they are variously called Schloemilch (or Schlömilch), Ferrié or Fessenden cells,[34-38] although as we shall see in the chapter concerned with thermal detectors,

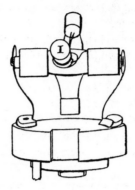

Fig. 4.7   *Multiple electrolytic detector introduced by the Telefunken Company in 1913*
Rotation brings any one of the three detectors into service
[*Electrical Review,* **72**, 1913, p.451]

it is problematical as to whether Fessenden's version actually worked electrolytically. Dr. Lee De Forest also claimed to have invented a detector of this sort to replace his 'responder', but the subsequent sharp exchange of letters in *The Electrician* between himself and Count Arco put him in his place rather firmly and established the prior claim of Schloemilch.[39]

The precise physical principle on which these detectors relied is really rather obscure, and modern books on electrochemistry are of little help since the phenomenon of producing depolarisation in this way has long descended into obscurity and become forgotten. Several engineers of the time, in particular Fessenden and Eccles,[40-42] inclined to the theory that the whole action was thermal in origin and was caused simply by changes in resistance due to heating of the minute layer of solution or gas around the small electrode where the current density was comparatively high. Another suggestion was that the layer of gas which formed at the fine point acted as an insulating dielectric so that capacitance existed in the cell. Change in the magnitude of this capacitance accompanied by flow of charge around the local circuit was thought to cause the audible click in the earphone.[43-44]

However, various workers,[45] and J.E. Ives in particular,[46] proved quite conclusively that it *was* due to a polarisation effect. It is per-haps worth noting in passing that Ives believed alkaline solutions to be best, but most people usually settled for diluted sulphuric or nitric acid. The results of Ives' experiments may be summarised briefly as follows:

(*a*) The device is not fully bidirectional in operation as it would be if it were thermal in origin.

(*b*) The fine point must be chemically inert if the action is to occur. If it is made of a metal which combines with oxygen, no detector action occurs.

(*c*) The device has a much higher electrical resistance than would be produced by a purely thermal effect.

(*d*) The sensitivity depends on the area of the small electrode in con-tact with the acid; the smaller the area, the better the action.

(*e*) It was a voltage-operated device, and not current operated as it would be if thermal.

(*f*) It provides detector action whatever the temperature of the liquid − even at boiling point.

(*g*) A 2·5% solution of hypophosphorous acid at 60° has a zero co-efficient of change of resistance with temperature, yet even with such an electrolyte the cell responds as normal.

(*h*) Platinum black absorbs oxygen. A thin layer of platinum black on the small electrode could not possibly affect the distribution of current in the electrolyte and could not disturb the action of a thermally operated device. Such a layer was found to stop the action completely, proving that a polarising film of gas was necess-ary for its operation.

Several people noticed that the asymmetrical electrode configuration was also able to act to some extent as a rectifier, i.e. it conducted current more easily in one direction than the other.[47−49] Ives was care-ful to point out that this was an effect of secondary importance and its efficacy as a detector in no way depended on it.

De Forest suggested that when the electrodes in an electrolytic cell were closely spaced, as in the Neugschwender mirror/slit detector des-cribed at the beginning of this chapter, it could act both as an anti-coherer with chains of particles and as a polarisation cell.[50] He pub-lished the interesting idealised curve of Fig. 4.8. This is a plot of current through the cell (vertical axis) against time (horizontal axis), and shows the response of the cell to six short bursts of oscillation, equivalent to six Morse dots. When the direct potential is applied initially, chains of particles form across the gap in the manner previously described and when the bridge is complete the current settles at a relatively high value as at *a* on the Figure. On application of the first burst of oscillation the chains rupture and the current decreases (as at point *b*). Polarisation effects then occur, the points of the ruptured chains acting as small electrodes causing the current to fall still further so that it eventually settles at point *c*. Arrival of the next burst destroys the polarisation as in the Schloemilch cell so that the current will rise to point *e*, falling again to *f* after the oscillation stops. This sequence repeats with the next pulse (*h* to *i*), but between the third and fourth bursts one or more of the chains succeed in reforming themselves so that the current rises to its high value again (at *k*). The sequence of events repeats for the last three bursts of oscillation, and when all oscillation has finally ceased it returns to the condition at *t* where the bridges are once again reformed.

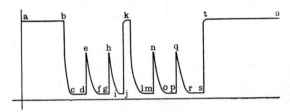

Fig. 4.8   *Idealised curve published by de Forest showing the response of an electrolytic cell with closely spaced electrodes to six bursts of oscillatory voltage, representing six Morse dots*
[*Electrician*, **54**, 1904, p.97]

There was also another slightly different method of using the phenomenon of polarisation.[51-52] If two disimilar metals are immersed in an electrolyte, a primary cell is formed. This produces an e.m.f. which can cause a current to flow around a closed circuit. After a little while, however, polarisation causes the e.m.f. of the cell to fall. An oscillatory voltage was again found to reduce the polarisation, producing a temporary increase in voltage which lasted until the oscillations ceased. The main advantage claimed for a cell of this sort was that it did not require a separate source of e.m.f. to work it; in other words, as well as being a detector it acted as its own battery. A telephone receiver of suitable resistance was connected across it; aerial and earth connections were made to the two electrodes, and current changes due to the arrival of signals could be heard. P. Jegou claimed to have used a cell of this type (with electrodes of a lead/tin alloy and a mercury/tin alloy immersed in sulphuric acid) at the wireless station at Ouessant (at the extreme western tip of Brittany), and to have received signals from Algiers, the Eiffel Tower and various other places. Operation of a cell in this way seems to have been moderately successful,[53] but several experimenters appear to have ended up by adding a local battery to 'back off' the steady current to a low value.

Another very convenient detector[54-56] of this sort was the Brown (or Hozier–Brown) peroxide detector, which is shown in diagrammatic form in Fig. 4.9, and also in Fig. 4.10. A moist pellet of lead peroxide

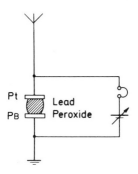

Fig. 4.9 *Principle of the Brown or Hozier-Brown peroxide detector*

Fig. 4.10 *Brown, or Hozier–Brown, peroxide detector*
[Science Museum photograph]

is pressed between two plates, one of lead and one of platinum. This constitutes a battery and is connected in series with a counter-e.m.f. supplied by another battery and a set of earphones. A small steady current flows around the circuit. An oscillatory voltage applied across the peroxide cell has the effect of 'stimulating chemical action' (as accounts in the literature describe it). This is probably just another way of referring to the depolarisation effect; anyhow the result is that the e.m.f. of the peroxide cell increases momentarily and the change in current around the circuit causes an audible click in the earphones. In the commercial form of the detector (Fig. 4.10) the pellet of lead peroxide was placed upon a platinum plate, and a blunt lead point mounted on a springy metal strip was pressed into it by adjustment of the knurled knob at the top. In the version shown, it was protected by a screw-on cover bearing the legend 'S.G. Brown's Patent Hertz Wave Detector', the lever at the side was a send/receive switch which allowed the detector to be disconnected from the aerial while the local transmitter was operating.

As stated earlier, there were a few detectors which, although electrolytic in action, did not fit very well into either of the two main categories. One such, due to Fessenden,[57-58] is shown in Fig. 4.11, which is reproduced from his patent specification of 1907. A tube (1) having

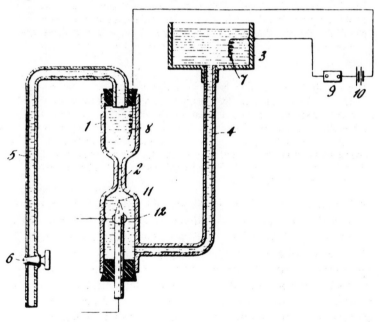

Fig. 4.11   *Fessenden's electrolytic detector using bubbles of released gas*
[British Patent 4714, 1907]

a constriction (2) at its centre is filled with dilute nitric acid. A very fine electrode (11) is surrounded by another electrode (12) made in the form of a ring. The high-frequency voltage oscillation is applied between these two electrodes. When an alternating voltage is applied in this way the gases released during one half cycle are recombined during the next so that no production of gases can be observed. According to Fessenden, if the area of the electrode is minute the gas does not cling to it but is immediately released into the liquid. It is not present during the next half cycle and therefore cannot recombine so that under these circumstances gas *is* produced. In this particular apparatus the gas rises into the constricted part of the tube where it may be observed with a microscope or recorded photographically. Its presence may also be detected electrically as follows. A local battery connected to the other two electrodes (7 and 8) will normally 'see' a certain resistance and a current will flow between them via the constriction. If a bubble of gas now appears in the constriction it will alter the resistance and the change can be observed by means of a galvanometer or telephone earpiece in the usual way. Fessenden states in the patent specification that

> In addition to the gas evolved directly by electrolysis, gas may be evolved by the heating effect of the alternating current waves on the fluid should the fluid contain a gas. This will also float upwards at the same time and will assist to produce an indication.

He goes on to say

> There will be a constantly flowing current between electrodes 7 and 8 and a constant evolution of gas which will produce a constant amount of obstruction in the restricted portion (2). When the current produced by the electromagnetic waves flows between terminals 11 and 12 additional gas will be evolved and a change will take place in the amount of obstruction in the portion 2. This change of obstruction will cause a change of current from the local battery (10).

It is very difficult to see how gas produced at electrodes 7 and 8 could possibly get up into the constricted part of the tube; such gases would float directly upwards out of harm's way. One suspects that this might be a statement included in the specification just in case some legal wrangle were to arise. In fact, it is very difficult to credit that the device could have worked at all in the manner stated. Is it really possible that when a radio-frequency alternating voltage is applied the gas can escape into the liquid so quickly that recombination is impossible in the next half cycle? Still, bearing in mind Fessenden's massive reputation as an inventor and engineer, one must assume that the apparatus worked somehow as a detector of radio waves, but it is beyond belief that this sort of thing could possibly have been used for telegraphy at any but the very slowest rates of Morse transmission. No doubt if the principle were sound it could have been used as a calling device to alert the operator and indicate to him that a message was about to be sent and that the receiving apparatus required his attention.

Another odd device, illustrated in Fig. 4.12 was that constructed by Zakrzewski.[59-61] Two jars A and B containing water are connected by a thin capillary tube silvered on the inside. Two fine platinum electrodes (*a* and *b*) are situated near the exits from this tube. Zakrzewski found that when air was pumped into one of the jars the

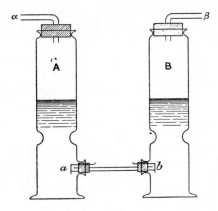

Fig. 4.12   *Zakrzewski's detector*
[*Electrician,* **46**, 1900, p.304, or *Phys. Zeitschr.,* 8 December 1900, p.146]

flow of water through the tube created a potential difference between the electrodes. The sense of the potential difference depended on the direction of flow and also, curiously, was reported as being dependent on the thickness of the silver coating on the tube. The inventor claims that it 'sometimes showed the behaviour of a coherer'. It seems likely that this was another form of the Schloemilch polarisation cell; we shall return to it shortly when we have looked at another of Fessenden's ideas which may help to explain its operation.

Fessenden suggested in one of his patents that it was advantageous to the sensitivity of electrolytic cells if the gas bubbles were removed physically from the small electrode leaving only a very thin film on it. To this end he devised the mechanical arrangements shown in Figs. 4.13 and 4.14.[62-63] In Fig. 4.13 the small point is set in a glass tube so that only a tiny area is exposed, and this tube is caused to move through the liquid by rotation of the pulley system on which it is mounted. 'The friction of the liquid against the exposed terminal will remove any gas which may form thereon.' In the alternative version of Fig. 4.14 the fine wire is again sealed in glass which is ground down flat so that a very small amount of platinum is exposed. This is pressed on

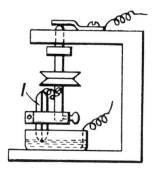

Fig. 4.13 *One of Fessenden's electrolytic detectors in which movement of the fine point (shown separately below) through the solution sweeps away the gas bubbles.*
[Blake, G.G.: *History of radio telegraphy and telephony* (Chapman & Hall, 1928), p.84]

a moist layer of cotton velvet, resting on another layer of absorbent material, all mounted on a metal disc. Rotation of this disc by the motor (15) again causes the gas to be swept away from the contact. Bubbling gas through the electrolytic cells or vibrating them was also claimed to improve their performance.[64]

There is really no way of assessing the efficacy of these various procedures without further experimental work, but could it possibly be that the motion of the water through the capillary tube in Zakrzewski's apparatus performed the same action of sweeping the bubbles away from the electrodes? It may be significant that he reported succes with the device only when the electrodes were very near the ends of the capillary tube, i.e. at the points where the water flowed with maximum velocity, and where the sweeping action would have been most effective.

E.R.W.D.—G

These then were the electrolytic detectors, methods of receiving which were reportedly reliable and sensitive when properly installed and adjusted. They saw service in several commercial systems during the first decade of the 20th century, although they were soon displaced by more convenient detectors.

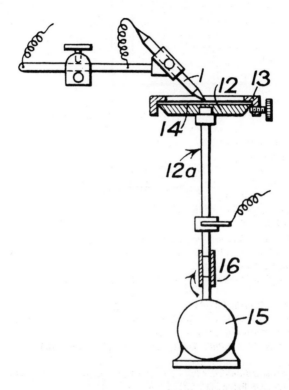

Fig. 4.14   *Another Fessenden detector where the fine point is pressed on a moving pad of absorbent material containing the liquid*
[Blake, G.G.: *History of radio telegraphy and telephony* (Chapman & Hall, 1928), p.84]

References

1   FLEMING, J.A.: *Proc. Roy. Soc.* (London), **71**, 1903, p.398
2   FLEMING, J.A.: *Principles of electric wave telegraphy and telephony* (Longmans, 3rd edn., 1916), p.491
3   MARX, E.: *Phys. Zeitsch.*, **2**, 1901, p.249
4   *Science Abstracts*, **4**, 1901 no. 1018, p.471
5   DESCHANEL, A.P.: *Elementary treatise on natural philosophy* (Blackie, 187?), Section 483, p.610
6   *Science Abstracts*, **2**, 1899, no. 567, p.232
7   MARX, E.: *Electrician*, **46**, 1901, p.611
8   NEUGSCHWENDER, A.: *Electrician*, **47**, 1901, p.396
9   MARX, E.: *Electrician*, **47**, 1901, p.434
10 *Scientific American*, Supplement, **52**, 1901, pp.21461 and 21512
11 DE FOREST, L. *Electrician*, **54**, 1904, p.94
12 HOWGRAVE-GRAHAM, R.P.: *Wireless telegraphy for amateurs* (Percival-Marshall, 1907 [?]), p.158
13 Reference 7
14 TOMMASINA, T.: *Comptes Rendus*, **128**, 1899, p.1092
15 .: *Electrician*, **43**, 1899, p.111
16 BLAKE, G.G.: *History of radio telegraphy and telephony* (Chapman and Hall, 1928), p.82
17 WALTER, L.H.: *Electrical Magazine*, **2**, 1904, pp. 384, 596
18 DE FOREST, L.: British Patent 10 452, 1902
19 *Electrician*, **52**, 1903, p.171
20 PIERCE, G.W.: *Principles of wireless telegraphy* (McGraw-Hill, 1910), p.204
21 PUPIN, M.I.: *Electrical World*, **34**, 1899, p.743
22 ERSKINE-MURRAY, J.: *Handbook of wireless telegraphy* (Crosby Lockwood, 4th Edn., 1913), p.127
23 WOLLASTON, W.H.: *Phil. Trans. Roy. Soc.* (London), **103**, Pt.1, 1813 p.114
24 *Encyclopaedia Britannica*, 11th Edn., 1910-11, Vol.28, see entry 'wire'
25 STONE, J.S.: US Patent 767 981, 1904
26 ECCLES, W.H.: *Wireless telegraphy and telephony* (Benn, 2nd Edn., 1918), p.288
27 COLLINS, A.F.: *Manual of wireless telegraphy* (Wiley/Chapman and Hall, 1906), p.52
28 DE BLONAY, R.: *Wireless World*, **2**, 1915, p.762
29 DE VALBREUZE, R.: *Notions générales sur la télégrahie sans fil* (Beranger,: Paris, 5th Edn., 1912), p.153
30 ZENNECK, J. and SEELIG, A.E.: *Wireless telegraphy* (McGraw-Hill, 1915), p.281
31 POWELL, S.M.: *Electrical Review* (London), **68**, 1911, pp.11 and 72
32 STONE, E.W.: *Elements of radio communication* (Van Nostrand, 1926), Section 518
33 *Electrical Review* (London), **72**, 1913, p.451
34 Reference 2, p.509ff

35   REICH, M.: *Phys. Zeitsch.,* 5, 1904, p.338
36   EICHORN, G.: *Wireless Telegraphy* (Griffin, 1906), p.81
37   SCHLOEMILCH, W.: *Electrician,* 52, 1903, p.250
38   FERRIÉ, G.: *Comptes Rendus,* 141, 1905, p.315
39   *Electrician,* 52, 1903/4, pp.240, 338, 382 and 620
40   TISSOT, C.: *Comptes Rendus,* 145, 1907, p.226
41   Reference 31
42   Reference 26, p.287
43   Reference 38
44   Reference 31
45   *Science Abstracts,* 7A, 1904, p.896, Abs. 2971

46   IVES, J.E.: *Electrical World and Engineer* (New York), 44, 1904, p.995
47   Reference 20
48   Reference 38
49   TORIKATA, W.: *Electrician,* 65, 1910, p.940
50   Reference 11
51   JEGOU, P.: *Electrical Review* (London), 67, 1910, p.170
52   Reference 17
53   Reference 37
54   Reference 2, p.511
55   Reference 22, p.191
56   BROWN, S.G.: US Patent 934 883, 1904, application granted 1909
57   FESSENDEN, R.A.: British Patent 4714, 1907
58   *Electrical Engineering,* 2, 1907, p.638
59   ZAKRZEWSKI, C.: *Physik. Zeitsch.* 2, 1900, p.146
60   *Electrician,* 46, 1900, p.304
61   *Harmsworth's wireless encyclopaedia* (Harmsworth 1923), Vol.2, p.836
62   Reference 16, p.84
63   FESSENDEN, R.A.: US Patent 1 022 539 application 1904, granted 1912
64   Reference 31, p.72

Chapter 5

Magnetic detectors

When a current flows in a wire, or in a coil of wire, a magnetic field is created. Various effects caused by such fields were used to reveal the presence of an oscillatory current in an aerial wire. Broadly speaking, these magnetic detectors may be considered to fall into two fairly well defined groups. In the first group, the magnetic field is caused to interact with another field thereby causing deflection of a needle, rotation of a mirror or some other mechanical movement. In the jargon of the time, these were often called 'electrodynamic' detectors. In the second group the oscillatory field caused by the aerial current changes the state of magnetisation of a piece of ferromagnetic material such as iron or steel. These were often referred to as 'hysteresis' detectors.

Of the electrodynamic detectors, perhaps the simplest was the Einthoven string galvanometer[1-3] which had originally been developed for purposes of electrocardiography. This was quite a well known instrument in which a very thin wire was held under tension between the poles of a strong magnet. Passage of a current through the wire gave rise to a mechanical force which deflected it. The deflection was viewed through a microscope, or in more elaborate installations an enlarged image of the wire was thrown on a screen. If an alternating current were used, then provided that its frequency was not too high (and the reader will recall that long waves of low frequency were commonly used). the wire was caused to vibrate so that a broad image was observed in the microscope. The arrangement could be tuned to some extent by varying the tension in the wire. Some workers seem to have used it directly in this form;[4] most, however, found it advantageous to include in the circuit some sort of rectifying device so that a unidirectional force was produced even when the frequency was quite high.[5-6] This made it much easier to observe and, if required, to record on a piece of moving film so that a permanent record of the incoming signal could be produced. Fig. 5.1 shows a record of a transatlantic signal recorded in this way.[7] An engineer used to more modern terminology might be tempted to raise here the objection that it was the diode which was doing the detecting and that the galvanometer was merely acting as an indicator. However, if we bear in mind the original basic definition of detection (the rendering visible of the presence of a radio wave) it is clear that it is the rectifier and galvanometer *together,* or the coherer and the telephone earpiece *together* , which are actually performing this function. Rectification as a *means* of detection took over so completely after the Great War of 1914-18 that it became commonplace to regard diode action as being the very essence of detection, but originally this was not so.

Returning to the Einthoven galvanometer; it seems to have been a very sensitive arrangement. Fleming, for example, reported that a current of 30 μA produced a deflection of several millimetres in the optical image of the wire thrown upon a screen[8], while Campbell-Swinton stated that movement was perceptible with currents as low as a ten-millionth of an ampere, or one billionth of a watt.[9]

An electrodynamic detector developed by Pierce[10-11] is illustrated in Fig. 5.2. A light metal disc (M) is suspended near a solenoid (C), the plane of the disc making an angle of 45° with the axis of the coil. An alternating current passing through the coil induces current in the disc which acts like a shorted single-turn transformer secondary winding. The primary and induced currents then experience mutual repulsion which causes the disc to rotate against the restoring torque of the suspension wire and to settle at some equilibrium angle depending on the magnitude of the current and the configuration of the nonuniform magnetic field. In Pierce's instrument some control of the sensitivity could be achieved by sliding the solenoid in and out in order to vary its distance from the disc. The disc also acted as a mirror and its deflection could be observed with the aid of a traditional light beam and scale arrangement. The first instruments using this principle of repulsion of primary and induced currents seem to have been invented independently by Fleming and by Elihu Thomson[12] and were used by them for the measurement of low-frequency alternating currents. Fessenden also patented a rather similar device[13-16] which is illustrated in Fig. 5.3. A small silver ring (8) was suspended between the two coils (7) with its plane at 45° to the axis of the coils. The coils were mutually coupled to the aerial circuit by the high-frequency transformer (11,12). The

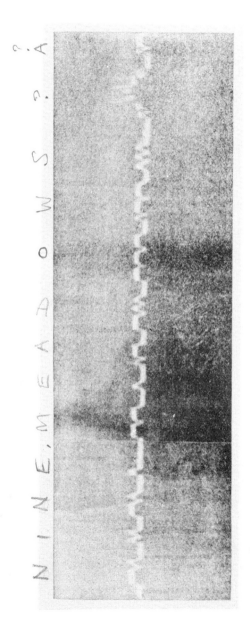

Fig. 5.1   *First published photographic record of a transatlantic message*
This message was transmitted from Glace Bay, Nova Scotia, and received at the Marconi station, Clifden, Ireland. The record was made by an Einthoven galvanometer actuated by the detector current [Erskine-Murray, J.: *Handbook of wireless telegraphy* (Crosby–Lockwood, 4th edn., 1913)]

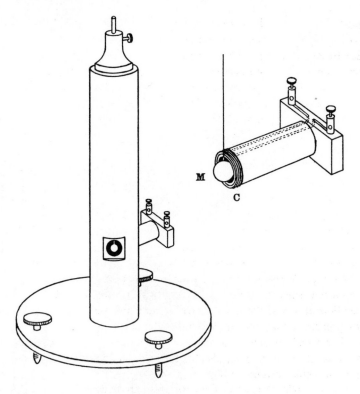

Fig. 5.2   *Pierce's electrodynamic detector*
[*Phys. Rev.*, **19**, September 1904, p.202]

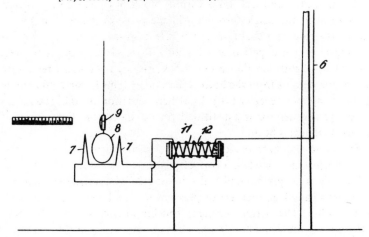

Fig. 5.3   *'Ring' detector invented by Fessenden*
[US Patent 706 736, 1902]

mirror (9) fastened on the suspension wire was used to observe the deflection. As far as can be determined from the literature, the version due to Pierce seems to have been the most refined of them, but according to Fleming it was not as sensitive as the best of the thermal detectors (q.v.) and it also had limitations of speed as far as the reception of Morse signals was concerned. Pierce stated in his paper that 'the period of swing one way is 3·4 seconds, and the mirror comes to rest after four oscillations'. Another difficulty which was experienced by him was that in the course of hanging the mirror (which was only 3 mm in diameter) on its quartz-fibre suspension it acquired an electrostatic charge which upset the deflection and caused the light spot to wander around the scale zero. This he cured by placing a piece of radioactive radium bromide in the case for a short while to neutralise the charge. This is reported to have cured the problem, the instrument remaining stable for many weeks after the treatment. Electrodynamic detectors, like the thermal detectors to be described later, had the useful property that they were able to provide an estimate of the root-mean-square value of the current since the force on the coil was independent of the current direction at any instant. As Pierce himself remarks in his paper, 'it is not presumed that such an instrument could be used as a receiver for wireless signals at any great distance, but without any great precaution in its use it proves to be applicable to the rapid and consistent measurement of feeble currents of high frequency'.

Another related type of detector was patented by Fessenden[17-19] and is illustrated in Fig. 5.4. Here the light metal ring (8) rests on two knife edges (13) and also on a carbon block (14). The aerial current flows through the single-turn coil (7) the plane of which is at 45° to that of the ring, and the transformer effect causes current induction in the ring so that it experiences a torsional force. This causes it to press more or less lightly on the carbon block, varying the contact resistance so that the current from the battery also varies. This is rendered audible by the telephone receiver (16). In a slight variation on this scheme an audio signal generator is substituted for the battery so that a note of varying loudness is heard.

The more elaborate version of Fig. 5.5 uses the two electrodynamic detectors just described in a sort of push–pull arrangement. An extra carbon block is placed under the other side of the ring so that the movement causes an increase of pressure on one block and a decrease on the other. The battery currents flowing in coils 16 and 16*a* cause deflection of the disc (17). It was claimed, not unreasonably, that this was more sensitive than the single-sided version of Fig. 5.4.

As far as can be determined, these *electrodynamic* detectors do not

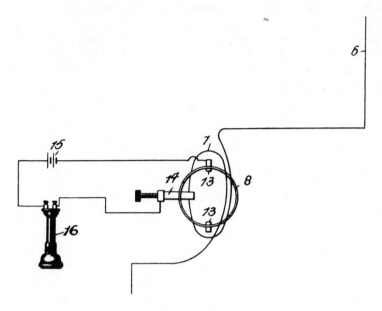

Fig. 5.4   *Fessenden's ring and carbon block detector*
[US Patent 706 736, 1902, British Patent 17 705, 1902]

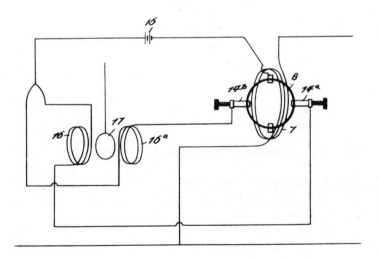

Fig. 5.5   *Combination of the detectors shown in Figs. 5.3 and 5.4*
[US Patent 706 736, 1902]

seem to have been used very much in practical communication systems. This cannot be said, however, about the *hysteresis* detectors, which were among the most useful and reliable of all the detectors and some of them were used extensively up to the time of the 1914–18 war. Before proceeding to a detailed study of these it will be as well to recall a few fundamental facts about the magnetisation of iron and steel. If a direct current flows through a coil wound around a specimen of magnetic material it creates a magnetic field, usually designated by the symbol *H*. This field causes magnetic flux, the density of which is represented by the symbol *B*. If the field is slowly increased in magnitude the flux in the specimen also increases and a graph may be drawn relating the two quantities. A typical curve for a ferromagnetic specimen is shown by the dotted line in Fig. 5.6. If the field is now decreased, the flux does not decrease along the same curve but along

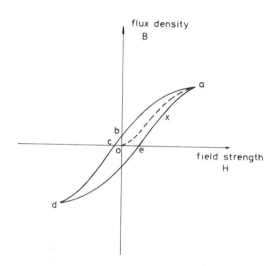

Fig. 5.6   *Magnetisation (B/H) loop for ferromagnetic material*

the path *ab*. Although the current and field may now be zero, a certain remanent flux represented by 0*b* remains in the specimen and it is necessary to reverse the field by amount 0*c* to reduce it to zero. This is the so-called 'coercive force'. If the current is increased further in the negative direction and then decreased to zero again in a cyclic manner the loop *abde* will be traced out repeatedly. This phenomenon was named 'hysteresis' by Professor Ewing, a pioneer in the study of magnetism.[20–21] It can be shown that the area enclosed by the *B/H* loop is a measure of the energy expended in taking the material through the

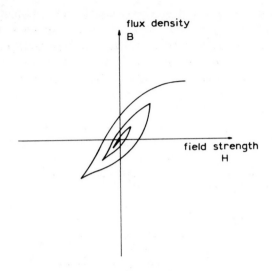

flux density
B

field strength
H

Fig. 5.7   *How demagnetisation is produced by subjecting the specimen to an alternating field of gradually decreasing magnitude*

magnetic cycle – the 'hysteresis loss'.

If the current flowing in the coil around the specimen happens to be an alternating current of steadily decaying amplitude the material will be taken through successively smaller and smaller loops, as shown in Fig. 5.7 so that eventually when the current has died away altogether the specimen will be left in an unmagnetised condition. This method of removing the magnetisation is often called 'degaussing', although strictly speaking the term should only be applied to a quite different method of cancelling magnetism by means of three mutually perpendicular coils.

In 1896 Rutherford constructed a detector of Hertzian waves which made use of this principle.[22-25] A bundle of magnetised steel needles or wires was placed at the centre of a solenoid and the ends of the solenoid coil were connected to the two sides of a Hertzian receiver in place of the usual spark-gap. About 20 steel wires were used, each about 1 cm long and 0·007 cm in diameter, separated from neighbouring wires by a layer of shellac varnish. A deflection magnetometer was used to monitor the state of magnetisation of the needles; (A deflection magnetometer is simply a little magnet suspended in the earth's magnetic field near the bundle of needles. The direction in which it points when it has settled is a measure of the relative strengths of the earth's field and that caused by the magnetism in the needles.) When a pulse of oscillatory current from the Hertzian receiver flowed through the coil the needles were partly demagnetised. Using this simple form of detector Rutherford was able to receive signals (of frequency about 45 MHz) over a distance of ¾ mile across the city of Cambridge. In a variant of this scheme, Dr. Huth suspended the needles themselves and their coil in the earth's magnetic field, the needles settling at an angle which depended on the relative strength of the magnetic forces and the restoring torque of the torsional suspension.[26]

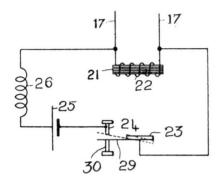

Fig. 5.8   *Wilson's detector with provision for automatic remagnetisation of the needles*
[British Patent 30 846, 1897]

These were obviously very clumsy and inconvenient devices requiring remagnetisation of the needles after the arrival of every pulse of radio-frequency oscillation. As an improvement, Wilson[27–28] constructed the system illustrated in Fig. 5.8. The pointer (29) attached to magneto-meter needle (23) normally rested against the stop (30), but after de-magnetisation of the needles (21) by a received signal the pointer would swing around until it made contact with the stop (24). Direct current flowing from the battery then remagnetised the needles in preparation for the next pulse. The actual apparatus used by Wilson is shown in Fig. 5.9.

A much more elaborate system of remagnetising was thought up by Fleming (Fig. 5.10).[29–30] In this detector, eight iron wires (26 SWG) 6 in. in length are put together to form a little bundle, and two coils are wound over them, one for magnetising, and one (the aerial coil) for demagnetising. The individual wires are separated by a coating of varnish. Seven or eight such bundles are inserted into the bobbins shown in Fig. 5.10. The coils wound on these bobbins are connected to the galvanometer G so that if there is any sudden change of flux the induced e.m.f. will cause deflection. A rotating shaft bears four fibre discs (1 to 4) each having a brass segment (shown in black) which makes contact with a springy brass strip (S_1 to S_4) once in each rota-

tion. The sequence of events is as follows. When contact is made with
S_1 a current flows from the battery via slip-ring S_5 through the magnetising coil so that the iron core is magnetised. During this part of the

Fig. 5.9   *Wilson's magnetic detector with arrangement for automatic remagnetisation of the steel needles after the arrival of the signal*
[Science Museum photograph]

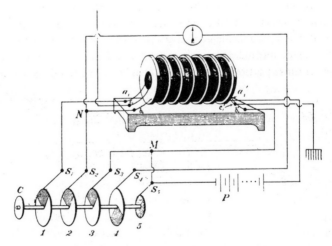

Fig. 5.10   *Arrangement constructed by Fleming where the needles are periodically remagnetised*
[Walter, *Electrical Magazine*, **4**, 1905, p.359]

operation discs 2 and 3 short out the aerial coil and disc 4 open circuits the galvanometer. Further rotation disconnects the magnetising current, and then removes the short from the aerial coil and reconnects the galvanometer. If a pulse of r.f. is received during the ensuing part of the cycle the wires are demagnetised and this is indicated on the galvanometer. The magnetism is restored during the next revolution of the shaft. If the incoming r.f. voltage is a continuous signal, and if the shaft is rotated at sufficient speed the galvanometer will settle at a steady reading. Fleming claimed that it was possible to make a quantitative measurement of the strength of the signal in this way.

Yet another variation on the demagnetisation theme was to use an adapted Poulsen Telegraphone.[31-32] This was a primitive sort of magnetic wire recorder. In this application a length of steel wire was uniformly magnetised and was then passed through a coil in which the oscillatory aerial currents flowed. The portion of the wire moving through the coil when oscillations were present was demagnetised so that when the wire recording was played back through the normal reproducing system the telegraphic signals were reproduced.[33-35] It was, in fact, simply making use of the 'erase' system found in modern tape recorders.

Fessenden's version of a scheme of this sort patented in 1908 is shown in Fig.5.11.[36] An iron or steel wire (41) passed first through the gap in the core of solenoid 45 which imposed a steady degree of magnetisation upon it. Solenoid 50 carried an audio-frequency alternating current generated by the vibrator (47) and this superimposed an amplitude modulation on the wire magnetisation. Coil 38 connected to the aerial acted as a demagnetiser so that the listener using the telephone (52) connected to search coil 51 heard an audio note during passage of those portions of the wire which had not been demagnetised.

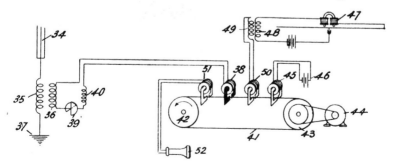

Fig. 5.11   *Fessenden's demagnetisation detector — a forerunner of modern tape recorder erase systems*
[British Patent 20 466, 1908]

This meant, of course, that the listener heard the note when dots and dashes in the signal were absent, and to avoid confusion Fessenden suggested sending an inverted signal where dots and dashes were represented by switching *off* the carrier.

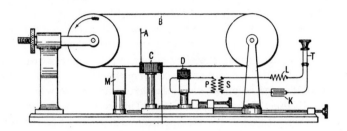

Fig. 5.12 *Shoemaker's demagnetising detector*
[Walter, *Technics*, **4**, August 1905, p.128]

Fig. 5.12 shows a rather earlier version due to Shoemaker.[37] In this, the magnetisation is simply produced by the permanent magnet (M); demagnetisation occurs in the aerial coil (C) and the changes in magnetisation are heard in the telephone via the search coil (D). Note that there is no audio-frequency modulation in this version or in the Poulsen telegraphone; the signal would have been heard merely as clicks at the beginnings and ends of the dots and dashes, where changes in flux occurred.

Two other magnetic detectors due to Fessenden are shown in Figs. 5.13 and 5.14. Although they are not strictly demagnetisation detectors, they do depend for their action on small changes in magnetisation due to superimposition of an r.f. field, and it will be convenient to consider them here before proceeding to a study of the more important cyclic hysteresis devices. The instrument of Fig. 5.13 dates from 1905, although the patent was not granted until several years later.[38] A carefully machined disc of brass (15), coated with a layer of magnetic material such as nickel or iron, is mounted on a shaft which is suspended magnetically by coil 19 and rotated by the permanent magnet (23) acting on disc 22. A very fine iron wire (11) — as thin as 0·001 of an inch in diameter, according to the inventor — suspended on quartz fibre (12) has its end (14) bent downwards so that it rests lightly on the disc. As the disc rotates, the friction causes the wire to deflect and to settle at a particular angle. The wire is linked mechanically to the delicately suspended pen of a siphon recorder (25) which makes a permanent record of any changes in position on a moving paper tape. The battery and potentiometer arrangement (26) adjusts the initial level of magnetisation and hence the friction, in effect setting the zero position of the pen. When aerial current passes through the coil (10) oscillatory changes are superimposed on the steady magnetisation. Since the force between the pen and the disc will depend on the square of the flux, the average force will increase slightly, being recorded by a movement of the pen. One could possibly have considered this detector to belong with the electro-dynamic devices, but since it does depend on the magnetisation characteristic of an iron specimen, it is probably better to classify it here.

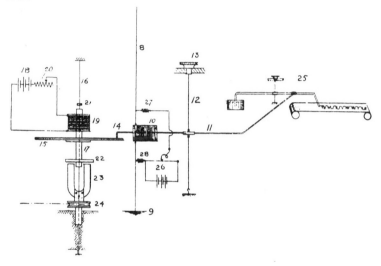

Fig. 5.13 *Frictional/magnetic detector by Fessenden*
[British Patent 6 008, 1909]

The other detector (Fig. 5.14) was patented in 1902.[39-41] A piece of iron or steel wire was held just above the poles of a strong permanent magnet (13). The pull of the magnet caused the wire to sag slightly. An oscillatory current passing through the wire changed its magnetic state to a small extent so that it rose and touched the contact (14) completing the circuit of battery (15) and bell or relay (16). Accounts of this were rather vague and it was not well explained in the contemporary literature. Unless the r.f. was of very low frequency or the wire was under great tension (in which case it would not have sagged very much), it is difficult to believe that the closing of the contact could have been caused by simple vibration of the wire due to the flow of current in the magnetic field as suggested by Fessenden in his patent. According to G.G.Blake's account referenced above, 'The oscillating currents through

E.R.W.D.—H

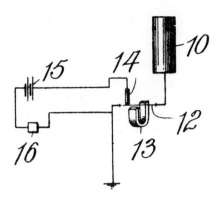

Fig. 5.14   *Fessenden's sagging-wire detector*
[British Patent 17 708, 1902]

the wire for a moment destroy the action of the magnetic field, and the wire springs up and touches a contact, momentarily closing a local circuit which includes a recording instrument'. Erskine-Murray offers a similar explanation in the second edition of his book *A handbook of wireless telegraphy* (1909). It may be that this was a rather peculiar manifestation of the hysteresis phenomena which will now be described.

Most of the hysteresis detectors were similar in general principle to that originally suggested by Wilson (occasionally ascribed to Tissot)[42-45] and developed by Marconi. It will be convenient to explain the action by reference to Marconi's model (patented in 1902)[46-48] which is shown in Figs. 5.15 and 5.16. A specimen of magnetic material in the form of a bundle of iron or steel wires *(a)* is situated between the poles of a permanent magnet. The magnet is rotated slowly about the vertical axis by means of a clockwork motor. The speed of rotation is by no means critical, but one revolution in every two seconds is said to have been suitable. As the magnet revolves, the specimen is magnetised alternately one way and then the other so that it is being cycled around its hysteresis loop. If a search coil having a large number of turns is now wound around the specimen its output will be found to be made up of two separate components. The first is a low-frequency variation which is simply proportional to the rate of change of flux, usually denoted $d\phi/dt$. This may be seen in the oscilloscope photograph of Fig. 5.18*a*. The other component is due to the so-called Barkhausen effect. The magnetic material can be considered to be made up of small areas known as 'domains', each one being a tiny magnet in its own right. As the applied field increases these move around one by

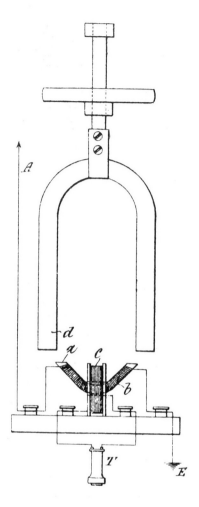

Fig. 5.15   *Marconi's first hysteresis detector*
[British Patent 10 245, 1902]

one to align themselves with that field, and as they do so the small flux change caused by the movement causes a small pulse of e.m.f. to be induced in the search coil so that a 'shushing' noise is heard in a telephone receiver connected across the coil. In the literature of the time this noise was often referred to as a 'breathing' sound. The present author has constructed a demonstration model of this detector and has found that this describes exactly the effect produced as the magnet rotates.

Fig. 5.16   *Marconi's first hysteresis detector, which used a rotating permanent magnet*
[Science Museum photograph]

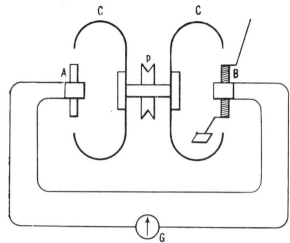

Fig. 5.17 *Tissot's 'back-to-back' system which cancels out the unwanted low-frequency output, enabling a galvanometer to be used to observe the arrival of a signal. C C are permanent magnets*
[Walter, *Electrical Magazine,* **4**, 1905, p.360]

Now a *BH* loop can be interpreted according to the following line of reasoning. If a sinusoidal voltage is applied to the X-plates of a cathode ray tube and the same voltage is also applied to the Y-plates then a straight line will be traced out on the screen. If, however, the phase of the voltage on the Y-plates is made to lag behind the X voltage this line opens out to form an ellipse. The line and the ellipse are examples of a series of traces known collectively as Lissajous' figures. We may regard the *BH* loop as being a somewhat distorted ellipse, indicating that the variation of flux lags behind the alternating field causing it. It is as though the little domains are reluctant to click around into alignment with the applied field and wait until a certain value of field is established before they are persuaded by the ever-increasing force to fall into line.

It was found that if an oscillatory current were passed through another coil wound over the specimen this somehow had the effect of jerking up the domains so that a large number of them would move around simultaneously. If, for example, the iron happened to be at point *x* of the cycle (Fig. 5.6) when the burst of oscillation occurred then the flux would tend suddenly to move nearer to the dotted line, which represents a 'no hysteresis' curve. The sharp change in flux caused by the domain movement caused a loud click to be heard in a telephone earpiece connected to the search coil.

Tissot's version of this detector[49] (Fig. 5.17) illustrates several interesting points. If the output of the search coil is observed by means of a galvanometer the sharp flux change caused by the arrival of a burst

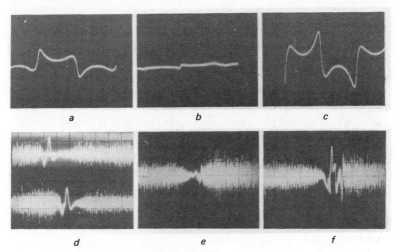

a   b   c

d   e   f

Fig. 5.18   *Experimental observations by the present author on Tissot's form of double hysteresis detector (see Fig. 5.17)*

(a)   Low-frequency voltage variation from one coil alone
(b)   From two coils connected in antiphase
(c)   From two coils in phase
(d)   Noise voltages from the two separate coils
(e)   Noise voltage from the two coils in antiphase
(f)   Noise voltage from the two coils in phase
       time scales: a-c 0·5s/cm; d-f 0·2s/cm

of r.f. in the aerial coil is rather difficult to spot because the galvanometer is constantly swinging to and fro due to the low-frequency $d\phi/dt$ voltage. To overcome this, Tissot connected two magnetic detectors back to back in such a way that the two low-frequency components were cancelled out. This is illustrated in Fig. 5.18 which shows some oscilloscope photographs taken by the present author. Photograph *a* shows the output of one search coil alone; photograph *b* shows the result of connecting two nominally identical coils in antiphase as in Tissot's arrangement, and, for comparison, photograph *c* shows the output voltage when the coils are connected in phase. As might be expected, the voltage here is double that in *a*. A galvanometer connected to the voltage *b* would clearly swing about much less that if it were connected to voltage *a*. In Tissot's apparatus, one specimen only was subjected to the oscillatory aerial current (coil B), producing a 'kick' in the galvanometer reading.

If the output from the search coil in these detectors was listened to by means of a telephone earpiece, the Barkhausen 'breathing' was found to be rather fatiguing, and tended to obscure the desired click. It might be thought that the connection of two coils back-to-back would also reduce this breathing, but this is not the case because the noises in the two coils are random and uncorrelated. Fig. 5.18*d* shows the outputs of the two coils after having been passed through a high-gain a.c. amplifier. The low-frequency component has been removed more or less entirely leaving just a small pulse at the point of maximum flux change, and the Barkhausen random noise. Photographs *e* and *f* show the result of connecting the coils in antiphase and in phase. In *e* the residual spike has been removed; in *f* it has been reinforced. However the noise voltage is the same in both cases. Theory predicts that in both cases the noise level would be $\sqrt{2}$ times that of a single coil, and the results illustrated are compatible with this prediction.

Thus, to summarise, the Tissot 'backing-off' procedure could cancel out the unwanted low-frequency swinging of the galvanometer, but could not reduce the breathing noises.

There was one great drawback to this type of hysteresis detector. It was only sensitive to the incoming signal when the flux in the specimen was in a state of change.[50] There were times during the revolution of the magnet when the material was starting to move back along the top or bottom of the hysteresis loop so that there was very little change in flux going on. As both Marconi and Tissot reported, the audible received signal was strongest when the pole of the magnet was approaching the specimen and weakest when it was moving away from it. This meant that a signal which happened to arrive during these periods of reduced sensitivity was likely to be missed altogether.

J.C.Balsillie[51] attempted to overcome this by using many pieces of magnetic material, each cycled through a *BH* loop, but at varying phases (Fig. 5.19). The description of his apparatus in *The Electrician* of 1910 reads as follows:

> Several small iron cores, each wound with a primary, through which the oscillations pass and collectively overwound with a secondary winding connected to a telephone are caused to rotate in a fixed magnetic field so that the flux passing through them is continually varying in intensity. Each core is in turn connected to the aerial when the magnetic cycle is at the critical point, and should oscillations be passing through the primary winding (i.e. the aerial coil) at this moment, the lag is instantaneously wiped out, the currents induced in the secondary winding (search coil) so actuating the telephone diaphragm.

Marconi's method[52] of overcoming the disadvantage of the insensitive periods is illustrated in Fig. 5.20*a* (Rutherford also claimed to have invented this particular method of operation).[53] Here a band of iron wires passes over two wooden pulleys which are rotated by a clockwork

Fig. 5.19   *Balsillie's hysteresis detector*
[Electrician, 64, 1910 p.514]

motor. The speed at which the wires moved was variously reported in the literature as 30 cm in 4 seconds, 1·6/cm second, and other widely differing figures; clearly much variation in speed was possible without affecting greatly the operation of the detector. Two horseshoe magnets are arranged as shown in Fig.5.20*b* with like poles adjacent so that any small portion of the wire passing beneath them is magnetised first in one direction, then the other. At the point where the magnetisation changes direction a coil is wound on a glass tube surrounding the wires, the ends of the coil being taken to aerial and to earth. A further search coil of many turns is wound on top of this and is connected to the telephone receiver. There is now a constant supply of iron which is rapidly changing its magnetic state passing under the coils so that there are no dead intervals during which a signal can be lost. Two versions of the detector are shown in Figs. 5.21 and 5.22. Marconi reported that the

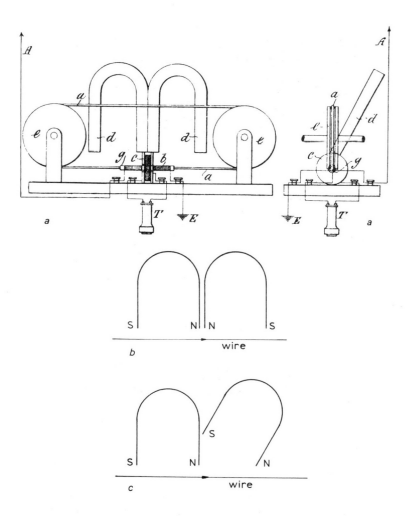

Fig. 5.20 *Marconi's second hysteresis detector*
(a)   Two views of the detector
(b)   and (c) Possible arrangements of the permanent magnets
[British Patent 10 245, 1902]

best results were obtained by using wires which had been stretched or twisted beyond the elastic limit before being made up into the continuous belt. The reader will no doubt note a certain superficial resemblance between this detector and that of Shoemaker (Fig. 5.12), but the principle of operation was quite different. As evidence of this we may note that Marconi's detector was symmetrical in that it would work whichever way the belt was moving, but Shoemaker's was not. It had to pass over the magnetising magnet first, then the demagnetising coil and lastly the sensing coil.

Fig. 5.21   *Marconi's moving-band hysteresis detector*
[Science Museum Photograph]

Fig. 5.22   *A Marconi hysteresis detector in use*
[Collins, A.F. 'Manual of Wireless telegraphy', (Wiley/Chapman & Hall, 1906) p.165]

As previously mentioned, operators found the Barkhausen noise rather annoying, and in the moving-band detector, since the iron was in a constantly changing state of flux there was a continuous hissing noise

Fig. 5.23   *Experimental observation of field configurations in Marconi moving-band hysteresis detector*
(a) corresponds to Fig. 5.20b
(b) corresponds to Fig. 5.20c

in the earphone which must have been very fatiguing to the ear during prolonged periods of listening. Many writers[54–57] suggested that this noise could be reduced by arranging the magnets as in Fig. 5.20c. It is not immediately apparent what this achieves, but the multiple exposure photographs prepared by the present author (Fig. 5.23) illustrate what happens. In these photographs the search-compass needle is moved along the line of the iron wire, and its direction indicates the direction of the magnetic field at each point. In photograph a, which shows the 'conventional' arrangement of magnets, the reversal of direction of magnetisation as the wire passes through the coils (between the two white dots) is evident. In the second arrangement however, a particular portion of the wire moving from the left will be magnetised in one direction when it is in between the poles of the first magnet, but as it progresses, it will be magnetised transversely, but will not have its direction of magnetisation actually reversed until it has passed way beyond the range of the search coil. It is, in effect, only subjected to half the reversal process. In terms of the magnetisation characteristic of Fig. 5.6; in the first case the longitudinal magnetisation changes from point a to point d, but in the second case only to point b. The Barkhausen noise is most pronounced on the steep part of the curve so it would naturally be reduced if magnetic reversal were not complete. It follows, of course, that the sensitivity would also be reduced in this arrangement, and this is verified by contemporary accounts. It seems to have been a practical engineer's compromise between decreasing noise on the one hand, and adequate (but reduced) sensitivity on the other.

In its commercial form (Fig. 5.21), Marconi's moving-band detector was found to be very satisfactory and troublefree and achieved considerable popularity. Its great virtue was that once it had been set up it remained in good adjustment for long periods without further attention. Its disadvantages were the need to wind up the clockwork motor at intervals and the fact that it was essentially a detector which needed an operator to listen to the signals and take them down by hand. It was not suitable for connection to a Morse inker which would automatically record the signals for decoding later. It was used in active service up to and including the Great War of 1914-18.

Various other people constructed their own versions of the hysteresis detector.[58-62] De Forest, for example,[63] devised the form shown in Fig. 5.24. When a piece of magnetic material is subjected to a high-frequency alternating magnetic field there is a 'skin effect' which means that the magnetic flux is confined to the outer layers of the material. With this in mind, he used a hollow cylindrical specimen and wound the aerial coil in two parallel sections along the inside and outside surfaces thereby doubling the flux by using both surfaces of the tube. His alternation of the magnetic field was produced simply by rotation of the bar magnet M.

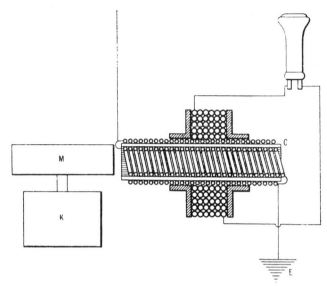

Fig. 5.24 *De Forest hysteresis detector*
[Walter, *Electrical Magazine*, **4**, 1905, p.360]

H.Bouasse[64] described the rotary form of Fig. 5.25. Here an iron or steel disc rotates between the poles of a permanent magnet in such a way that any small area of material is magnetised one way and then the other. Aerial and search coils are wound as shown and function in the normal manner.

An extremely complicated version due to Fessenden,[65] that most prolific of inventors (he is said to have had over five hundred patents to his credit) is shown in Fig. 5.26. The coils (drawn in dotted lines) wound on a steel disc are connected to two alternating voltage sources (top right) having a phase difference of 90°. The coils are wound in such a way that a rotating magnetic field is set up; thus every particle of steel within the influence of the coils is taken through its magnetic cycle. Another winding (drawn in full line) wound right around the disc in toroidal fashion carries the aerial current, and two more wound across the centre of the disc at right angles act as search coils. These are connected to two solenoids situated on the limbs of a horse-shoe shaped core, the poles of which are very near a thin iron diaphragm forming a telephone earpiece. Capacitors in these circuits reduce the effects of the low-frequency e.m.f.s induced by the rotating field itself, but permit the passage of the pulse caused by the sharp flux change on arrival of a signal. This moves the diaphragm, causing an audible click.

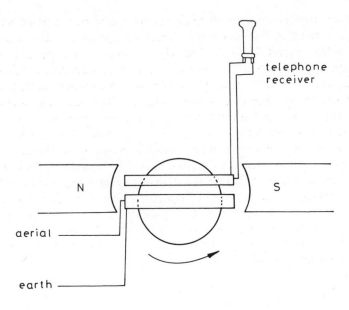

Fig. 5.25   *Rotary form of hysteresis detector described by Bouasse*

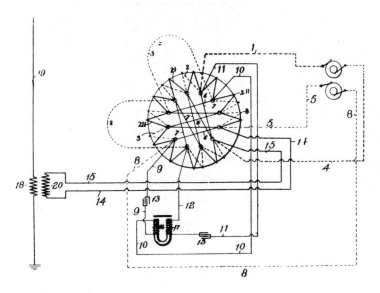

Fig. 5.26   *Fessenden hysteresis detector which used a two-phase coil configura-*
tion to produce a rotating magnetic field in the disc-shaped specimen
[British Patent 26 553, 1902]

Since the operation of all these hysteresis detectors is based on a sudden change of flux produced in a magnetic specimen, both Marconi and Wilson suggested that this flux might be arranged to act directly on a diaphragm, thereby doing away with the need for a search coil. Wilson, in the patent of 1902, included a design for a hysteresis detector built into a telephone earpiece as shown in Fig. 5.27. Two bent iron rods (14 and 15) are arranged to have small gaps between them at the points where they approach the diaphragm (16). The coils (17) are connected to a source of alternating current in order to cycle the iron through the *BH* loop, and further coils (C) are connected to aerial and earth as usual. The sudden change of flux on reception of a signal causes movement in the diaphragm. The permanent magnet (18) and screw (19) may be used to set the diaphragm at the optimum distance from the gaps. Another diagram in the same patent shows how the diaphragm may be replaced with an armature and contact so that a Morse inker or sounding coil might be operated.

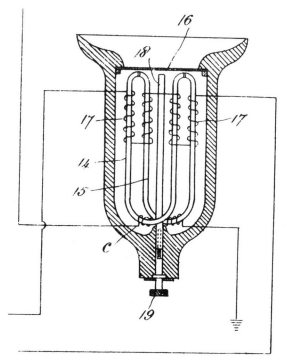

Fig. 5.27   *Hysteresis detector constructed by Wilson in the form of a telephone earpiece*
[British Patent 14 829, 1902]

It was natural that experimenters who were trying to use hysteresis effects should seek to adapt existing apparatus which was already being used to study these effects. One of the best known of these hysteresis testers was that invented by Professor J.A.Ewing.[66-68] It is illustrated in Fig. 5.28. A C-shaped permanent magnet is mounted so that it can swing vertically on knife edges and a bar of the magnetic material under test is rotated in the space between the poles. Because of the hysteresis lag already discussed the direction of maximum flux in the specimen always lags behind the direction of maximum field by an angle which is usually called the hysteresis angle. If the magnet were free to revolve it would follow the specimen around. This is the basis of the hysteresis motor which is used today in specialist applications. In Ewing's tester there is a restraining torque due to gravity acting on the magnet so that

Fig. 5.28   *Ewing hysteresis tester*
*[JIEE, **24**, 1895, p.400]*

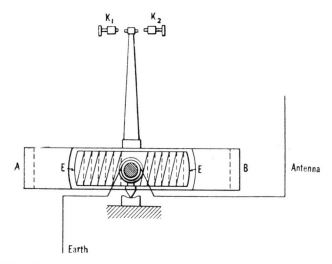

Fig. 5.29   *Adaptation by Peukert of Ewing's hysteresis tester*
[Walter, *Electrical Magazine,* **4**, 1905, p.361]

it simply swings out of the vertical. A pointer moving over a scale indicates the extent of the swing and it can be shown that this is proportional to the area of the hysteresis loop of the specimen. The magnet is also supplied with a damping vane dipping down into an oil chamber in order to smooth out the action and provide a steady reading. Variations on this arrangement are possible; for example, the magnet could be rotated and this would produce a torque on the specimen. Again, if the whole thing were to be mounted horizontally a coiled spring could be used instead of gravity to provide the restraining torque.

Several people wound an aerial coil on the specimen itself. Normally the pointer would settle at a reading on the scale, but the passage of an oscillatory current through the coil would disturb the magnetic domains causing a change of deflection. A version of this due to Peukert[69-70] is shown in Fig. 5.29. The specimen (E) rotates; the magnet with poles A and B swivels around it. Slip-rings enable the aerial current to pass through the coil wound on the specimen, and in this particular case the change in deflection of the magnet makes one of the contacts K so that a relay or inker may be operated by a subsidiary battery circuit.

Walter and Ewing constructed a detector of this sort in 1904, but found that the results were rather feeble and disappointing.[71-73] They then decided to pass the oscillatory current through the specimen itself, thereby producing a circular field within the material. Rutherford, in one of his earlier papers, had suggested operating in this way and it may have been his work which gave Walter and Ewing the idea. They found that the effect was much larger, but they also obtained the surprising result that the apparatus indicated an *increase* in hysteresis loss when a short burst of oscillations occurred. They then went on to produce the form of hysteresis detector shown in Fig. 5.30. The specimen in this case is a silk-covered steel wire wound on a bone bobbin. Contact is made to this wire at the bottom by an end dipping into mercury or, as in the version shown, by means of a small slip-ring. Contact at the top is through a small hair-spring which also serves to provide a restraining torque against the rotation of the bobbin. Various arrangements of windings were used, but a noninductively wound bifilar winding (an insulated steel wire wound back on itself) permitted the greatest oscillatory current to pass, and produced the most sensitive instrument.

M M are the wedge-shaped pole pieces of an electromagnet which rotates around the specimen at about 5 to 8 revolutions per second, being supplied with current from the large slip-rings at the bottom. Finally, the bobbin is totally immersed in oil 'to fortify the insulation and to give the damping effect necessary to steady the deflection due to the drag of the rotating magnet'. A mirror mounted on the bobbin shaft enables its rotation to be observed. As previously stated, Walter and Ewing found that oscillatory current from the aerial flowing through the coiled specimen caused a change in deflection in a direction indicative of an increase in hysteresis loss. Walter subsequently published details of an even more refined instrument of this sort[74] which confirmed the result, and it was also noticed by Professor Arnó[75-79] who used the apparatus of Fig. 5.31. In this a disc (D) made up of iron and steel particles set in paraffin wax is suspended in a rotating magnetic field created by the three-phase coils A, B and C. The disc experiences a drag due to hysteresis lag which would normally cause it to follow the rotation of the field, but an identical disc is suspended beneath, mounted on the same spindle. This, too, is situated in a rotating field, but its coils (A', B', and C') are connected in such a way as to produce field rotation in the opposite direction. The result is that the spindle does not move. With this apparatus, Arnó observed that when an oscillatory current was passed through another coil wound over the top disc its lag was disturbed and the spindle moved in a direction which, again, indicated that hysteresis loss in the top disc had increased.

As previously remarked, this increase was totally unexpected and not well understood. When discussing their instrument, Walter and

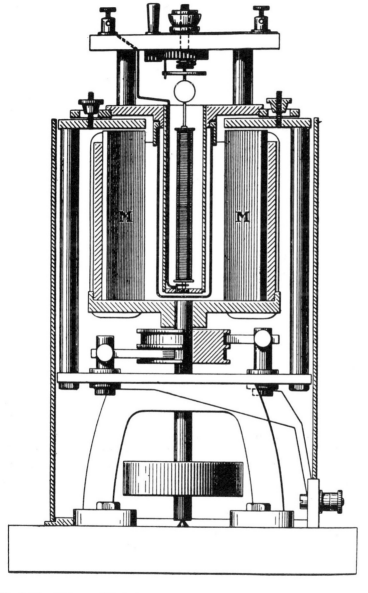

Fig. 5.30 *Walter and Ewing hysteresis detector*
[*Proc. Roy. Soc.,* **73**, 1904, p.120]

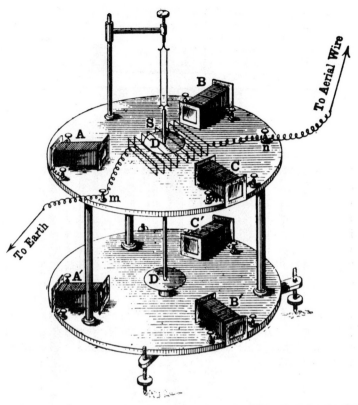

Fig. 5.31   *Arno's apparatus using rotating magnetic fields set up by the three-phase coils ABC and A'B'C'*
[*Electrician,* **53**, 1904, p.270]

Ewing had ventured to suggest that[80]

> the augmentation of hysteresis is interesting and unlooked for. It is probably to be ascribed to this, that the oscillatory circular magnetisation facilitates the longitudinal magnetising process, enabling the steel to take up a much larger magnetisation at each reversal than it would otherwise take and thus indirectly augmenting the hysteresis to such an extent that the direct influence of the oscillations in reducing it is overpowered.

This brief explanation is not easy to grasp at first reading, but what they said essentially was that

> (a) When a cyclic magnetising force is applied in the usual way the material is taken through its hysteresis curve. Owing to the fact that a circular magnetic field is also present (due to the oscillating current flowing through the specimen

from end to end) this helps the *magnetising* process so that
a given peak value of the main magnetic field causes a much
greater value of flux. In these circumstances, a much larger
BH loop than usual is traced out.

(*b*) The oscillations also have the effect of shaking up the
domains in the usual way and reducing the lag, but the loop
itself is so much larger than it would have been without the
oscillations that this normal effect is swamped. The overall
effect registered is thus an increase in the area of the loop.

Several people attempted to resolve this question, perhaps the most
thorough investigation being that of James Russell[81] in 1905. His work
was aimed at determining the effects of superimposing codirectional or
transverse oscillating fields on the main field. It would be impossible to
deal with every aspect of his very detailed paper here; rather let us
select some of his results to illustrate the point at issue. He first made
clear the important distinction that whereas the Marconi detectors
made use of a single instantaneous change of flux at one point in the
BH loop, the types based on the Ewing tester were essentially integra-
ting instruments which smoothed out and averaged the combined effects
of oscillatory pulses over several cycles of the loop. The curves of Fig.
5.32 are taken from his paper. The full-line loop shows the normal
hysteresis curve measured in the absence of oscillatory currents; the
dotted and stepped curve shows the loop produced when several bursts
of oscillation are superimposed throughout the course of one cycle. The

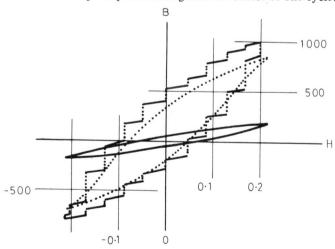

Fig. 5.32   *Hysteresis curves published by James Russell showing the effect of*
bursts of high-frequency oscillation (stepped curve) and continuous
oscillation (dotted curve)
The full curve shows the hysteresis loop in the absence of oscillations
[*Proc. Roy. Soc.* (Edinburgh,) **26**, 1905/6, p.42]

dotted curve is produced by keeping the oscillations on continuously. It will be seen that the second and third loops enclose much larger areas than the first, indicating that the energy loss per cycle has indeed increased. Russell also concluded that it does not really matter whether the oscillatory field is codirectional with the main field (as in the Marconi detectors) or transverse (as in the Walter and Ewing detector); in addition, that the overall effect depends on the absolute level of the magnetisation. With relatively small values of main cyclic field an increase in hysteresis loss would be expected, but with high fields a decrease would result. He predicted that at some intermediate value of field there would be 'uncertain results or no results at all from magnetic detectors of this kind'. This prediction was subsequently confirmed practically by Professor Arno.[82-84]

Before leaving the subject of the Ewing type of hysteresis detector it is perhaps worth looking at one due to Fessenden[85] (Fig. 5.33). In this case an annular disc of magnetic material (22) is situated near a magnet (21) mounted on swivel bearings, and it is rotated by a clockwork motor (25). In the normal course of events the magnet would rotate with the disc, but it is restrained from so doing by a string connected to diaphragm 23. When rotation is regular the whole thing will settle with the string in a state of tension. Aerial current flowing via slip-rings through a coil wound toroidally on the disc will result in the usual phenomena, causing a momentary change in the tension and an audible sound from the diaphragm.

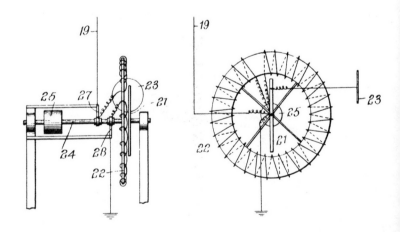

Fig. 5.33   *Two views of Fessenden's directly audible hysteresis detector*
[British Patent 26 553, 1902]

There was yet another variation on the theme of altering the hysteresis characteristic. It was known to be affected by heating and mechanical stress,[86-87] and the aforementioned James Russell also published another very detailed paper dealing with these effects.[88] It has already been remarked that the performance of the Marconi iron-belt detector was improved if the iron had previously been stretched beyond the elastic limit, and in describing the detector of Fig. 5.26 Fessenden had specified that the steel disc should be placed in a state of mechanical stress by clamping it at the edge and applying a force at the centre (or vice-versa). Wilson, in one of his patents dealing with the production of a low-frequency cyclic field by means of a solenoid, made provision for the magnetic specimen to be heated by two small flames. Apparently it was most sensitive when heated to a temperature just below the Curie point where magnetisation disappears.[89] He also reported, in an article written in 1902, that tension or torsion increased the magnetic effects in a bundle or wires. The Italian A. Sella[90-92] showed that 'a detector of less delicacy may be made by applying the discovery that the magnetic state of iron is sensitive to electric waves when the hysteretic cycle is due to elastic deformation instead of to variations in an external field'. His apparatus was improved by L. Tieri[93-96] who took a bundle of iron wires and soldered them all together at their ends (Fig. 5.34). A glass tube was slipped over the bundle and one end was firmly clamped

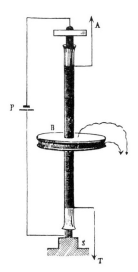

Fig. 5.34   *Tieri's torsional detector*
[*Accad. Naz. Lincei Atti,* **15**, 1906, 4th February, p,165]

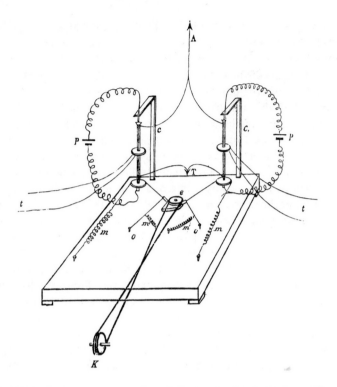

Fig. 5.35   *Pulley arrangement for twisting two torsional detectors with a mutual 90° phase difference*
[*Accad. Naz. Lincei Atti,* **15** 1906, 4th February, p. 167]

so that the bundle was held vertically. A single-layer coil was wound on the glass tube, being connected to aerial and earth (Terre) in the usual way. A lever was attached to the free end of the wire bundle so that it could be twisted one way and then the other. He claimed also to have improved the performance of Sella's detector by allowing a direct current to pass through the bundle at the same time as it was being twisted and untwisted. A search coil was also wound on the glass tube and was connected to a telephone earpiece. The arrival of an oscillating aerial current when the twisting was taking place resulted in a loud click being heard. A core made of nickel wires was also found to be very satisfactory. This device suffered from the same sort of disadvantage as the first of the Marconi hysteresis detectors in that the core was only sensitive when undergoing torsion. Twice in each cycle of twisting and untwisting its ability to detect incoming signals decreased very markedly. To improve this he constructed the version shown in

Fig. 5.35. This had two cores twisted by a series of levers and pulleys in such a way that when one core was at a dead point the other was at maximum sensitivity. Tieri himself claimed in his paper that this was more sensitive than the Marconi detector and that with this apparatus radio-telegrams were well received. However, in writing the *Science Abstract*,[97] L.H.Walter felt it necessary for some reason to add the phrase 'apparently from a few metres distance' which, if true, must have been rather disappointing after all that effort!

A variant on this was the detector invented by A.G.Rossi.[98-100]

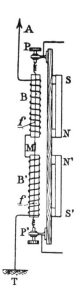

Fig. 5.36 *Rossi's torsional detector*
[de Valbreuze, R.: *Notions générales sur la télégraphie sans fil*
L'Elairage Electrique, 6th edn., 1914), p.124]

this employed a phenomenon known as the Wiedemann effect. According to the *Encyclopaedic dictionary of physics*[101] this effect concerns 'The twisting of a ferromagnetic under the simultaneous action of circular and longitudinal magnetic fields. In a rod, for example, the resulting lines of force are helices about the axis of the rod and any magnetostrictive effect causes the rod to twist'.

Rossi suspended an iron or nickel wire vertically under tension, and fixed two bar-magnets parallel and near to it with the two north poles near the centre of the wire (Fig. 5.36). A small mirror was also attached to the wire at its mid point. When an audio-frequency alternating current (at about 42 Hz) was passed through the wire the Wiedemann effect produced torsional vibrations in the wire which could be observed by a light beam reflected from the mirror. By adjustment of the tension in the wire a condition of resonance could be achieved in which quite a large amplitude of vibration occurred. Two coils were wound around the two halves of the wire, these coils being wound in contrary directions. The passage of a high-frequency aerial current through these coils was found to cause a decrease in amplitude of vibration, and hence a shortening of the strip of light cast upon a screen by reflection from the mirror. According to one account, a single burst of r.f. produced very little effect, but the instrument was very sensitive to a series of bursts, especially if the burst frequency happened to be the same as the frequency of oscillation of the wire. It was also claimed that one could dispense with the alternating current; it seems that a received train of bursts could produce vibration by itself if the bursts occurred at resonance frequency.

Interesting and unusual as these torsional detectors were, there is no evidence to suggest that any instrument employing such effects ever became practically useful receivers for wireless telegraphy signals. Of all the magnetic detectors, it was the Marconi iron-band hysteresis detector which was outstandingly successful.

References

1 FLEMING, J.A.: *Principles of electric wave telegraphy and telephony* (Longmans, Green, 3rd Edn., 1916), p.713
2 ZENNECK, J., and SEELIG, A.E.: *Wireless telegraphy* (McGraw-Hill 1915), p.294
3 MARCHANT, W.H.: *Wireless telegraphy* (Whittaker, 1914), p.115
4 EDELMAN, P.E.: *Experimental wireless stations* (Edelman, Minneapolis, 3rd Edn., 1915/16), p.173
5 *Electrician,* 63, 1909, p.908
6 ROUSSEL, J.: *Wireless for the amateur* (Constable, 1923), p.114
7 ERSKINE–MURRAY, J.: *Handbook of wireless telegraphy* (Crosby–Lockwood, 4th edn., 1913), frontispiece
8 Reference 1
9 CAMPBELL–SWINTON, A.A.: *Electrician,* 72, 1914, p.687
10 PIERCE, G.W.: *Physical Review,* 19, 1904, p.196
11 Reference 1, p.527
12 FLEMING, J.A.: *Electrician,* 18, 1887, p.561
13 FESSENDEN, R.A.: US Patent 706 736, 1902
14 SEWALL, C.H.: *Wireless telegraphy* (Crosby–Lockwood, 1904), p.162
15 BOUASSE, H.: *Oscillations éléctrique* (Delagrave, Paris, 1921), p.280

16 BLAKE, G.G.: History of radio telegraphy and telephony (Chapman and Hall, 1928, p.74
17 FESSENDEN, R.A.: British Patent 17 705, 1902
18 References 13, 14 (p.159) and 16
19 JENTSCH, O.: *Telegraphie und Telephonie ohne draht* (Springer, 1904), p.178
20 EWING, J.A.: *Proc. Roy. Soc.* (London), **38**, 1885, p.58
21 EWING, J.A.: *Electrician*, **14**, 1885, p.479
22 RUTHERFORD, E.: *Phil. Trans.* (A), **189**, 1897, p.1
23 RUTHERFORD, E.: *Electrician*, **49**, 1902, p.562
24 RUTHERFORD, E.: *Proc. Roy. Soc.* (London), **60**, 1896, p.184
25 Reference 15, p.300
26 WALTER, L.H.: *Electrical Magazine*, **4**, 1905, p.359
27 WILSON, E., and EVANS, C.J.: British Patent 30846, 1897
28 WILSON, E.: *Electrician*, **51**, 1903, p.330
29 FLEMING, J.A.: *Proc. Roy. Soc.*, (London), **71**, 1903, p.398
30 Reference 1, p.498
31 *Electrician*, **53**, 1904, p.269
32 Reference 16, p.212
33 SIMON, H.T., and REICH, M.: *Electrical Magazine*, **1**, 1904, p.560
34 *Science Abstracts*, **7B**, 1904, no. 1148, p.426
35 Reference 26
36 FESSENDEN, R.A.: British Patent 20 466, 1908
37 WALTER, L.H.: *Technics* **4**, 1905, p.127
38 FESSENDEN, R.A.: British Patent 6 008, 1909
39 FESSENDEN, R.A.: British Patent 17 708 1902
40 Reference 16
41 Reference 14, p.163
42 TURPAIN, A.: *La télégraphie sans fil* (Gauthier–Villars, 1908), p.31
43 Reference 15, p.301
44 MAZZOTTO, D.: *Wireless telegraphy and telephony* (Whittaker, 1906), p.192
45 TISSOT, C.: *Comptes Rendus*, **136**, 1903 p.361
46 MARCONI, G.: British Patent 10 245, 1902
47 MARCONI, G.: *Electrician*, **49**, 1902, p.520
48 MARCONI, G.: *Proc. Roy. Soc.* (**London**), **70**, 1902, p.341
49 Reference 26
50 ECCLES, W.H.: *Electrician*, **57**, 1906, p.742
51 BALSILLIE, J.G.: *Electrician*, **64**, 1910, p.512
52 References 46–48
53 Reference 23
54 HARRIS, P.W.: *The maintenance of wireless telegraph apparatus* (Wireless Press, 1917), p.65
55 ECCLES, W.H.: *Wireless telegraphy and telephony* (Benn, 2nd end., 1918), p.285
56 Reference 3, p.72
57 HAWKHEAD, J.C., and DOWSETT, H.M.: *Handbook of technical instruction for wireless telegraphists* (Wireless Press, 2nd edn., 1915), pp.176–7
58 Reference 37
59 WILSON, E.: British Patent 14 829, 1902

60   WILSON, E.: *Electrician*, **49**, 1902, p.917
61   Reference 45
62   Reference 44, p.179
63   Reference 26
64   Reference 15, p.317
65   FESSENDEN, R.A.: British Patent 26 553, 1902
66   EWING, J.A.: *JIEE*, **24**, 1895, p.398
67   EWING, J.A.: *Electrician*, **43**, 1899, p.43
68   FLEMING, J.A.: *Handbook for the electrical laboratory and testing room. Vol. II (Electrician*, 1903), p.478ff
69   BERTHIER, A.: *La téléphonie et la télégraphie sans fil* (Desfarges, Paris, 1908), p.191
70   Reference 26
71   WALTER, L.H., and EWING, J.A.: *Proc. Roy. Soc.* (London), **73**, 1904, p.120
72   WALTER, L.H.: *Electrical Magazine*, **1**, 1904, p.277
73   Reference 15, p.317
74   WALTER, L.H.: *Proc. Roy. Soc.* (London), **77A**, 1906, p.538
75   ARNÓ, R.: *Electrician*, **53**, 1904, p.269
76   *Electrical Magazine*, **2**, 1904, p.166
77   *Science Abstracts*, (A) **8**, 1905, no.1048, p.346
78   Reference 26
79   Reference 44, p.196
80   Reference 71
81   RUSSELL, J.: *Proc. Roy. Soc.* (Edinburgh), **26**, 1905/6, p.33
82   ARNÓ, R.: *Electrician*, **55**, 1905, p.469, 450, 588
83   *Science Abstracts*, (A) **8**, 1905, no. 2289, p.714
84   *Science Abstracts*, (A) **9**, 1906, no. 1617, p.482
85   FESSENDEN, R.A.: British Patent 26 553, 1902
86   EWING, J.A., and COWAN, G.C.: *Electrician*, **21**, 1888, p.83
87   EWING, J.A.: *Electrician*, **14**, 1885, p.479
88   RUSSELL, J.: *Proc. Roy. Soc.* (Edinburgh) **29**, 1908/9, pp. 1, 38
89   WILSON, E.: *Electrician*, **49**, 1902, p.917
90   *Science Abstracts*, (A) **7**, 1904, no. 2685, p.806
91   MAURAIN, C.: *J. de Physique*, **6**, 1907, p.380
92   Reference 26
93   *Science Abstracts*, (A) **9**, 1906, no. 767, p.224
94   *Science Abstracts*, (A) **9**, 1906, no. 1618, p.483
95   TIERI, L.: Acad Naz dei Lincei: Atti Rendiconti, Classe di Sci Fis Math e Naturali, **15**, 1906, p.164
96   *Science Abstracts*, (A) **9**, 1906, no. 1616, p.482
97   *Science Abstracts*, (A) **9**, 1906, no. 767, p.224
98   ROSSI, A.G.: *Nuovo Cimento*, **15**, 1908, p.63
99   *Science Abstracts*, (A) **11**, 1908, no. 687, p.252
100 ROSSI, A.G.: *Electrician*, **64**, 1909, p.116
101 THEWLIS, J.(Ed.): *Encyclopaedic dictionary of physics* (Pergamon, 1962), Vol.7, p.761

Chapter 6

Thin-film and capillary detectors

These two types of detector will be considered together because they both make use of phenomena occurring in fluids, although in many ways those in the first group really belong with the coherers. In the literature of the time these were often called simply 'liquid coherers'. The reader may recall that in most of the coherers described in Chapter 3 two conductors were kept apart by a thin insulating film usually of the metal oxide, but sometimes by the sulphide or by some other substance such as resin. The insulating coating was broken down by the incoming signal allowing the metals to make good contact. The principle of the thin-film detectors was the same except that a film of oil or some other fluid replaced the oxide.

In 1897 Rollo Appleyard published a short paper entitled 'Liquid coherers and mobile conductors' in which he described the effects of applying an electrical potential difference to mercury/oil mixtures.[1-2] In one of his experiments, mercury and paraffin in equal proportions were shaken together in a tube until the mercury had been broken up into fine droplets and dispersed evenly throughout the paraffin. It then exhibited a very high resistance when probed by two electrodes inserted at the ends of the tube. If the electrodes were then connected to a source of voltage, the droplets suddenly coalesced so that a resistance of only a few ohms resulted. He observed that the same effect could be obtained by holding the tube near a Hertzian oscillator or by passing the spark directly through the mixture. The paper goes on to describe various other fascinating phenomena where droplets of mercury were seen to send out small tentacles to touch one another, or were observed to crawl along the tube like little caterpillars.

Campanile and Di Ciommo continued these experiments and built a somewhat more practical form of mercury/oil coherer.[3-4] Strips of glass were cemented onto a glass plate in order to make a rectangular trough which was then filled with a mixture of mercury and Vaseline, stirred up into drops. This was found to act like a normal coherer on application of an oscillating voltage, and further investigation revealed that conducting chains of coalesced drops were formed. Further stirring was needed to decohere it. Several experimenters constructed liquid coherers consisting of two drops of mercury separated by a film of oil; such a detector, for example, appears in one of Ernest Wilson's patent specifications (Fig. 6.1).[5] The present author remembers only too well his own attempts to chase blobs of mercury around the school physics bench with pieces of filter paper, and one can only marvel at the patience shown by these investigators. One technique of doing it, apparently, was to cut a pool of mercury in half with a knife previously coated with paraffin.[6] The thin coating of paraffin thus formed was sufficient to keep the two halves apart until the application of a voltage between them caused them to relapse into a single drop again. One author[7] quotes some calculations made by Lord Rayleigh in which he showed that if one considered two surfaces separated by a film 10^{-7} cm in thickness, maintained at a potential difference of one volt, then the force of electrostatic attraction between them would be 650 lb/square inch. The coalescence seemed to be caused by this force literally squeezing out the film of oil, and in support of this it was said that it took some little time to achieve it — the coalescence was not instantaneous. Although all these experiments must have been most interesting to watch, one need hardly comment on the unsuitability of such phenomena for use in practical signalling systems.

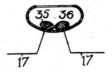

Fig. 6.1   *Mercury-drop and oil-film coherer described by Wilson and Evans* [British Patent 30 846, 1897]

Lodge, Muirhead and Robinson, however, produced a much more useful device which they arrived at by stages.[6-11] Their first liquid detector consisted simply of an iron point dipping down through a layer of oil into a pool of mercury. The surface tension of the oil ensured that there was a thin insulating film formed between the point

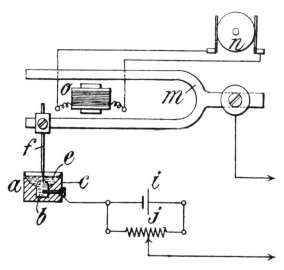

Fig. 6.2   *Oil film detector constructed by Lodge and Muirhead*
The contact point (*f*) was vibrated by the tuning fork to ensure
prompt reforming of the oil film
[British Patent 13 521, 1902]

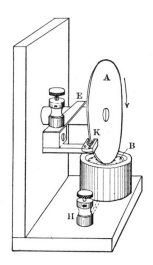

Fig. 6.3   *The Lodge–Muirhead–Robinson disc coherer*
[Pierce, G.W.: *Principles of wireless telegraphy* (McGraw–Hill, 1910),
p.144]

and the mercury, but when the aerial voltage was applied between them the film was ruptured, allowing them to come into contact. It was restored to the sensitive state by withdrawing the point and allowing it to fall back again so that the film was reformed. In an improved version of this (Fig. 6.2) the point was fixed to a constantly vibrating tuning fork so that the film was reformed automatically as soon as it had ruptured.

The final form of their device was the very successful disc or wheel coherer seen in Fig. 6.3. A thin steel disc having a smooth sharp edge (A) is rotated by a clockwork motor. It just touches the top of a pool of mercury on which floats the film of oil, this being contained in the cup (B). In this arrangement the rotation of the disc sweeps a thin film of oil between the disc and the mercury so that there is no electrical contact. On application of the aerial voltage between the disc and the mercury the film is broken down so that contact is established. As soon as oscillations cease, the rotation sweeps the oil around again so that the high-resistance condition is re-established. Several of the authors quoted in the references gave very detailed accounts as to how the device was best set up initially. First of all, the contact between the wire and the mercury at the bottom of the cup had to be very good in order to avoid untoward coherence effects at this point which might affect the proper action. Monkton described how to clean it should it have become contaminated with oil.[12] The contact wire had to be removed and heated to red heat, then plunged into some mercury in order to form a surface amalgam. The edge of the disc had to be free of dust and without notches or imperfections of any sort. It was recommended that a spring-loaded piece of cork or leather should bear upon the edge of the rotating disc in order to keep it clean at all times (K in Fig. 6.3).

The speed of rotation was not critical, but it ought not to be too slow or else the film would not have time to reform completely between the dots and dashes of the message. Howgrave–Graham, for example, specified a speed of half a revolution per second for a 7/16 inch diameter wheel.[13] When setting it up initially, he also advised the user to give 'promiscuous dot and dash signals' to ensure that it was operating correctly. We should point out that the word 'promiscuous' in those days was used to mean 'of various kinds mixed together'.

The sensitivity of the Lodge – Muirhead – Robinson wheel detector was very dependent on the thickness of the oil film, and in the commercial model the mercury cup could be moved up and down by means of a fine screw so that optimum conditions could be achieved (Fig. 6.4).[14] When properly set up a potential difference of 0·1 V was sufficient to cause rupture of the film and a direct bias voltage of this order was often applied to increase its sensitivity to the incoming oscillations. Several writers stressed the fact that if it were biased in this way it was essential to make the wheel positive with respect to the mercury, although it is not easy to see why this should have been the case. As previously stated, this was a very successful detector which saw practical service for a number of years.

Fig. 6.4   *Two versions of the Lodge–Muirhead–Robinson disc coherer*
[Science Museum photographs]

E.R.W.D.—K

Rather more bizarre were the devices known collectively as Fessenden's frictional receivers.[15-17] Two of these are illustrated in Fig. 6.5a and b. A strip of very thin aluminium or gold foil (2) suspended from a bar (6) just touches the edge of a smooth revolving wheel or cylinder. In Fig.6.5, the discs rotate anticlockwise. In the normal course of events, the friction between the wheel and the foil would be sufficient to drag the foil downwards so that the pivoted bar would be tilted and the pointer would move up along the scale (7). A fine jet of fluid – and Fessenden recommended the use of milk for this purpose in one of his patents – is allowed to trickle between the foil and the wheel, lubricating the contact and allowing the bar to remain horizontal. (The reader will no doubt wish to ask why milk was chosen for this purpose. No reason was given; one must just assume that it happened to be both handy and suitable!) The foil strip was connected to the aerial and the wheel to earth, and if a voltage appeared the lubricating film broke down allowing the foil to touch the wheel again so that the pointer moved up the scale. In the version of Fig. 6.5b the foil was fixed directly to a membrane (3) so that the movements of the foil would be made audible.

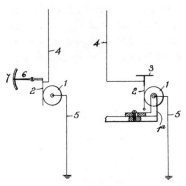

Fig. 6.5   *Fessenden frictional receivers*
[US Patent 1 042 778, 1912]

Another arrangement is shown in elevation and in plan in Fig. 6.6,[18] and in this version the foil (148) is wrapped around a cylinder (146) which also dips into a trough of lubricating fluid. The foil is connected by a fine piece of thread to a siphon recorder, which is simply a fine glass tube with a jet at one end pivoted loosely at the point labelled '150'. The rear end of this tube dips into a reservoir of ink which is transferred to the jet by siphonic action. Movement of the foil due to frictional changes also moves the tube, and the signal is recorded on a strip of moving paper.

In the version of Fig. 6.7 a disc (158) rotates in a horizontal plane and a small piece of mica covered with gold leaf on the back (160) rests on it, held in place by threads. The incoming signal applied between disc and gold leaf again varies the friction and results in movement of the mica plate. The threads (164) are connected to a vertical bar (94) situated in a trough of carbon granules (lower diagram). The sides of the trough are connected through a centre-tapped transformer to a battery, the other pole of which is connected to the bar. As the bar moves on its pivot the resistance of the granules changes in a manner similar to that in a carbon microphone, so that unbalanced currents flow in the transformer and a signal is heard in a telephone receiver connected to the secondary winding. The use of a lubricating fluid is not specified for this device; it was probably used dry. Readers who are familiar with the history of loudspeakers may recognise here something rather similar to the Johnsen–Rahbeck effect which was used as a means of magnifying the movement of a loudspeaker diaphragm. This early use of the 'push–pull' principle with a centre-tapped transformer is also worthy of note.

In yet another version, Fessenden specified the use of two discs operating in a mechanical push–pull arrangement. However, it must be admitted that there is no evidence to suggest that any of these frictional receivers were ever more than laboratory curiosities.

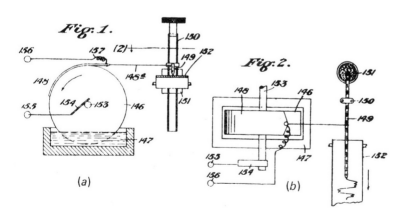

Fig. 6.6   *Two views of the Fessenden direct-writing frictional detector*
[British Patent 11 154, 1910]
(*a*) side view
(*b*) plan view

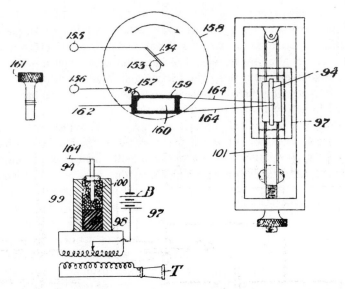

Fig. 6.7   *Fessenden frictional detector coupled to a carbon-granule microphone*
[British Patent 11 154, 1910]

Another type of thin-film detector was invented by
L.H.Walter,[19-20] and is shown in Fig. 6.8*a*. A thin capillary tube (on
the left of the diagram) dips down into a vessel containing a layer of
distilled water on top of mercury. Inside the capillary there is a wire
which comes to within 1/60th of an inch of the end of the tube. When
the end of the capillary tube is pushed down about 1/20th of an inch
into the mercury the mercury surface is depressed owing to surface
tension and it cannot enter the tube and make contact with the wire.
This is shown in Fig. 6.8*b*. Wire and mercury are therefore separated by
a thin layer of distilled water, which is a poor conductor. The mercury
is earthed, and when a voltage from the aerial (11) is applied to the thin
wire the surface tension changes in such a way as to allow the mercury
to enter the tube, as in Fig. 6.8*c* and to come into contact with the wire.
There then follows an interesting series of events. The current which
flows around the circuit of cell 9 operates a relay (10) which in turn
energises the solenoid (8). This jerks the balanced beam on which the
capillary is suspended so that it is pulled up out of the mercury. The
solenoid is thus de-energised and the tube falls back so that the insulating film is reformed and it is ready to receive the next voltage pulse
from the aerial. The message can be read from the movement of the
beam, or by means of an indicating instrument in the solenoid circuit.

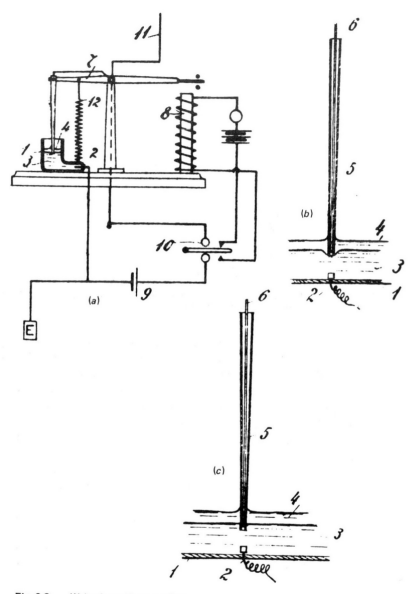

Fig. 6.8 *Walter's capillary receiver*

(a) The circuit, showing the arrangement for restoration to the high-resistance state

(b) The high-resistance state with distilled water separating the wire (6) and the mercury (3)

(c) The low-resistance state with wire and mercury in contact

[British Patent 17 111, 1902]

Alternative forms of restoration suggested were:

(i)   A blast of air blown down the tube and bubbling out through the jet.
(ii) Movement downwards of the vessel, as opposed to upwards by the tube.
(iii)   Movement of the glass sheath only, leaving the thin wire stationary.

Loss of water due to evaporation seems to have been a problem and it was suggested that the whole receiver should be enclosed in a glass case, a small reservoir for purposes of topping-up being provided.

Two further detectors may be considered to belong in this section where we are discussing two conductors separated by a thin film of fluid, although the fluid in these is actually air. The first of them is the device usually called the 'Lodge ball-coherer'. Oliver Lodge[21-22] was experimenting with a lightning protection device consisting of a small spark-gap which would conduct the lightning safely to earth. He found that when the two sides of the gap were separated by a very small amount, the occurrence of the flash would cause them to come together by electrostatic attraction and stick until they were mechanically disturbed. He went on to construct a radio detector consisting of two metal balls in close proximity which came into actual metallic contact when the aerial voltage was applied between them. This device is often classified simply as a coherer in the literature and is included with the filings coherers and similar devices. However, since its action seems to depend on the squeezing out of a thin film of air the present author has chosen to include it here. Lodge himself was none too certain as to whether the separation was by air, by oxide or (as seems very likely), by both.

The last thin-film device which will be described goes by the rather jolly name of 'Boys's bouncing jets'. Over a period of some twenty years Lord Rayleigh had carried out a series of experiments on jets of water.[23-26] He showed that if two jets of acidulated water collided at a very small angle then they would not coalesce, but would bounce apart again, being separated by a thin film of air, and continue on their way as two separate jets. If, however, an electrical potential were applied between them they joined together and carried on as a single stream of water. C.V. Boys[27-29] made use of this in the detector whose principle of operation is shown in Fig. 6.9. Two fine jets of water, a and b, issuing from two glass tubes are shown in the normal uncoalesced position. When the aerial/earth potential was applied between them they combined in the manner just described and in so doing struck a little

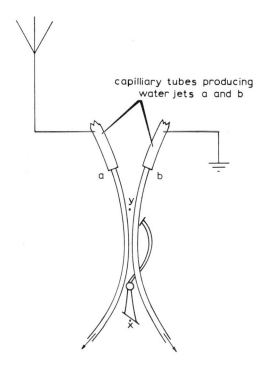

Fig. 6.9 *Principle of operation of Boys's bouncing jet detector*

pivoted lever, the turned-up end of which was situated at the point x.
Because the obstruction was at an angle to the jet it caused the lever to
move so that its other end, suitably shaped, was brought across to the
point y thereby separating the jets again. The lever then returned to its
former position under the influence of a light spring. When it moved
the lever was also caused to close a pair of contacts so that an inker or
some other recording devices could provide a permanent record of the
signals. Another variant used a solenoid-operated lever to separate the
jets after application of the aerial voltage. It was reported that a potential difference of about 1 V was necessary to cause the jets to join
together, and so application of a bias voltage of about this amount
improved the sensitivity. For the apparatus to operate reliably it was
necessary to ensure that the jets were produced by a constant head of
water, and the precautions necessary to achieve this condition in
practice may be seen in Fig. 6.10 which shows a drawing from Boys'
patent specification. The water levels in the two jars were kept constant
by the floats (*b*) which operated the valves (*f*) controlling the flow of

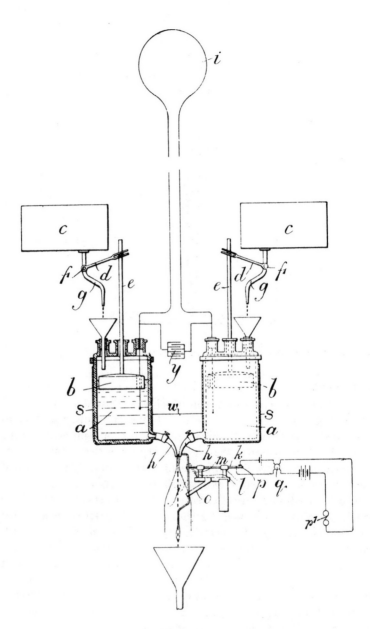

Fig. 6.10 *Boys's bouncing jet detector; the practical apparatus*
[British Patent 13 828, 1905]

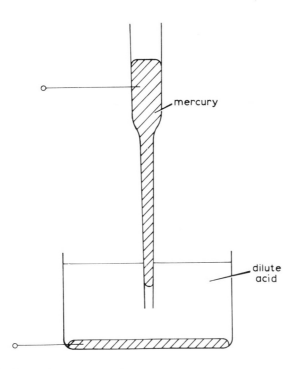

Fig. 6.11   *Simple Lippmann capillary electrometer*

water from the reservoirs (*c*). It will also be seen that a loop aerial (*i*) is depicted in this case and that a slightly different arrangement of the lever from that previously described is shown in this particular version. Water from the coalesced jets strikes the lever (*j*) and forms a droplet on the end of it, the weight of which overbalances it slightly and causes the necessary movement.

So much then for the thin-film detectors; let us now turn our attention to the capillary detectors which were based on the electrometer invented by Gabriel Lippmann in 1874.[30-32] The simplest form of the Lippmann electrometer is illustrated in Fig. 6.11. A glass tube, drawn out to form a thin capillary, dips into a vessel containing dilute acid. The capillary tube contains mercury which, because of its surface tension, does not run out of the bottom but remains some little distance up the tube. If a potential difference is now applied between the mercury in the tube and some lying on the bottom of the acid container there is a change in surface tension which results in movement of the mercury/acid interface in the tube. The physical explanation for this change in surface tension is very complex and not well understood (by the present author anyhow!) but it undoubtedly makes a very sensitive instrument and if the movement of the interface is observed with a microscope it may be used for measurement of potential differences of 2mV or less. In fact, in his original papers Lippmann stated that it could be used to measure the potential difference of 1/1000th, and to estimate 1/10 000th of that of a Daniell cell (which gives 1·08 V on open circuit).

It must be admitted that the literature poses something of a problem here. It was clearly a device operated by a direct voltage and therefore one would expect to have to use some sort of rectifier to make it respond to high-frequency oscillatory signals. Some experimenters, however, appear to have been able to use it without a rectifier. The reader will, perhaps, bear this in mind until we have described some of the actual apparatus they used and we shall return to consideration of this point a little later.

T.J. Burch[33] made a very detailed study of the properties of the Lippmann electrometer and gave very precise instructions as to how it should be constructed if reliable results were to be achieved. Absolute cleanliness was of the utmost importance; any dirt would cause the mercury to stick or catch at some point in the capillary. There were many ways of observing the movement, the most obvious being to use a microscope as mentioned previously. Another method was to connect the capillary tube to a manometer arrangement which could be used to bring the mercury back to the same mark at each measurement, the voltage being determined by observing the position of the reservoir of the manometer. Another very popular arrangement was to project the image of the capillary onto a large screen[34] by means of a system of lenses or to record the movement photographically.[35] J.Roussel used the rather more elaborate capillary system of Fig. 6.12*a* and projected it onto a screen by means of the optical apparatus of Fig. 6.12*b*.[36] In his book *Wireless for the amateur* (which was published as late as 1923) he explained how he arranged for the image to be projected onto a screen at a distance of 6 m, and reported that the signal from the Eiffel Tower, when applied to the capillary caused the image to move by 1·8m. G.Vanni also stated that 'e.m.f.'s of a few 10 000ths of a volt can be placed in evidence on the screen'. These figures give some idea of the sesitivities achieved by the capillary electrometer detectors.

The reader may wonder at this point why the Walter capillary detector of Fig. 6.8 is not included here, since it involves movement of a mercury/water interface. The reason is that in Walter's apparatus the fluid (pure water) was chosen for its insulating properties, whereas the Lippmann devices are concerned with the interface of two conducting

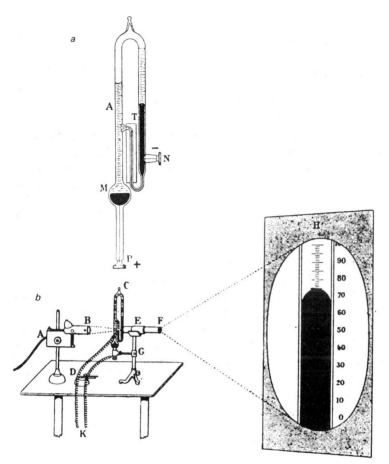

Fig. 6.12 *Capillary receiver showing (a) the capillary tube and (b) the optical apparatus for casting an image of the mercury/acid interface on a screen*
[Roussel, J.: Wireless for the amateur (Constable 1923), p.257/9]

fluids, mercury and acidulated water. The distinction is, admittedly, rather a fine one, and one would not wish to be too pedantic about it.

Antoine Bréguet[37] made use of the Lippmann effect to construct the novel form of telephone receiver shown in Fig. 6.13*a*. (Although this is not a radio detector and so, strictly speaking, has no place in this book it is nevertheless a good illustration of the application of the phenomenon, and is closely related to the Plecher receiver, which will be described shortly.) If potential differences are applied between contacts L and L_1 the mercury will move. This causes movement of the membrane stretched across the wider part of the tube and it can be heard by an ear applied to the top of the funnel. A rather more elaborate version constructed by Mr. Charles Lever of Cheshire[38] in 1885 is shown in Fig. 6.13*b*. Here the mercury *m* in the glass tube G is in contact with sulphuric acid in the enclosed chamber S. A light piston or rod P rests on the surface of the mercury, and by adjustment of the reservoir A its upper surface is brought into contact with a metal diaphragm D. Movement of the mercury/acid interface is thus communicated to the diaphragm. It is interesting to note that Bréguet and Lever claimed to have used identical instruments as telephone transmitters, claiming that just as a change in voltage causes movement of the mercury, so does movement of the mercury cause a potential difference to appear; i.e. the effect is reversible. Lippmann himself commented on this reversibility in his 1874 paper.

Plecher adapted this principle to the reception of radio signals;[39-40] his receiver is shown in Fig. 6.14. The section on the left has two capillary tubes of the sort just described (to double the effect, one assumes). A membrane is stretched above the mercury (Hg) in the reservoir and the space above is in connection with two flexible tubes which may be inserted into the ears so that the pressure changes occurring as the Morse signals arrive may be heard directly. There were several noteworthy features of Plecher's apparatus, not the least being the composition of the electrolyte which consisted of a solution of potassium cyanide with 1% silver cyanide and 10% potassium hydrate. Quite why Plecher chose to use this lethal mixture is not made clear; one

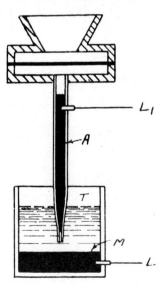

Fig.6.13*a*

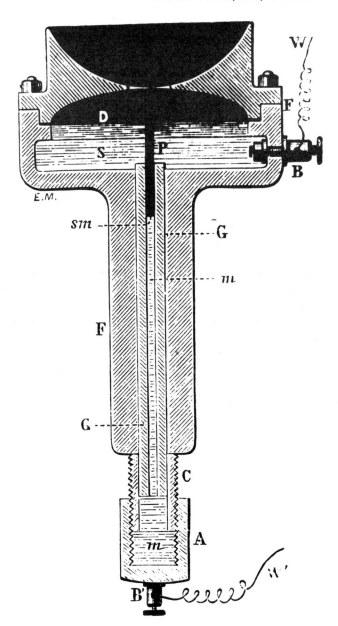

Fig. 6.13*b* *Two capillary telephone receivers*
(*a*) By Brėguet, (*b*) Properly engineered version due to Lever
[Blake, G.G.: *History of radio telegraphy and telephony* (Chapman &
Hall, 1928), p.20: *La Lumiėre Électrique,* **16**, 1885, p.428]

suspects that he tried all sorts of things, adding a pinch of this and a pinch of that until best results were achieved. Another feature of his apparatus was that a second capillary was provided (on the right of the diagram). This was arranged so that on commencement of signals the movement of the mercury completed the bell circuit and woke up the slumbering operator.

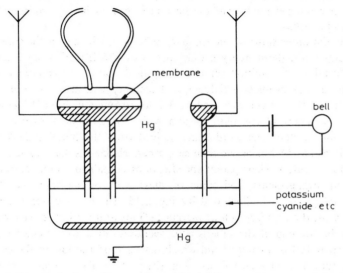

Fig. 6.14   *Plecher's capillary receiver with subsidiary receiver to sound the alarm bell and summon the operator*

Returning now to the question of the need for rectification before using the capillary electrometer; J.Roussel, in his account of it, makes it very clear that a rectifier was necessary and stated quite specifically that it had to be used with a crystal. In the accounts of Plecher's receiver, however, there is no mention of the need for rectification; the impression is given that the capillary was connected directly between aerial and earth. J.Erskine-Murray,[41] in his description of an electro-capillary receiver, simply states that it may be used as a receiver of 'jigs' (the rather attractive name he gave to bursts of damped oscillations). An account given in the *Electrical Review* of 1904[42] says 'the recorder is also applicable to wireless telegraphy, being used in series with an ordinary coherer — the special feature of this application is that when practically no current is allowed to pass through the coherer, as in this case, the latter is self-decohering'. No details of the particular type of coherer are given, but could it perhaps have been introducing an element of rectification, thereby enabling the capillary electrometer to work? It may simply be that the writers of these accounts had not fully understood what was going on; we shall have occasion to return to this question yet again in connection with some apparatus invented by Armstrong and Orling, but one is tempted to wonder in passing whether there might possibly have been some self rectification at the acid/ mercury junction, or some asymmetry in the surface tension effects which produced movement of the interface even in the absence of a crystal rectifier.

If the dimensions of the capillary tube are such that in the absence of any e.m.f. the mercury is only just restrained from flowing out, the application of a voltage can cause a critical change in the surface tension so that droplets will form, and mercury will drip into the acid solution. This effect was used by J.T.Armstrong and Axel Orling in a d.c. earth-current signalling system which came to be known as the 'Armorl' system, the word being derived from the initial syllables of their names,[43] but as we shall see, others claimed to have used it for radio reception. Their patent specifications contained a number of different arrangements, and some of these will now be described. The general principle is illustrated by Fig. 6.15. The lower vessel contains the acid; the capillary full of mercury (*f*) dips into it in the usual way, but the top end of the tube is bent over and dips into the taller vessel which is full of mercury, and which makes good the loss in the capillary tube as the mercury drips out at the bottom. The complete apparatus is shown in Fig. 6.16. When a voltage is applied between electrode (*i*) in the capillary and electrode (*j*) in the acid a drop of mercury falls, and it tilts the light lever (*k*) which completes the contact (*o*) which in turn operates the inking recorder (*q*). Naturally the release or nonrelease of a drop depends critically on several factors – the diameter of the orifice, for example – and also on the pressure of the mercury in the capillary tube. It was reckoned to be essential to keep the level of mercury in the taller vessel constant, and to this end the reservoir (*r*) was introduced. When the level starts to fall, it uncovers the orifice (*u*) and this allows a bubble of air to enter the reservoir so that some mercury runs out and tops up the level in the vessel.

In the version of Fig. 6.17 the mercury and the acid are arranged in a concentric manner. The acid is contained in the left-hand reservoir and the mercury in the container at the top. When no voltage is applied, the acid enters the central capillary tube and no mercury flows out, but when a signal appears the changes in surface tension allow a drop of mercury to fall in the usual way. The drop momentarily connects the two electrodes under the jet as it falls and this operates a recording

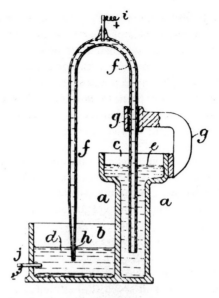

Fig. 6.15   *Illustrating the principle of the 'Armorl' mercury drop detector*
[British Patent 21 981, 1901]

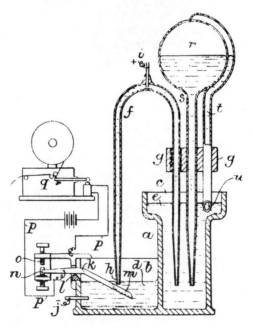

Fig. 6.16   *More elaborate version of the 'Armorl' receiver*
[British Patent 21 981, 1901]

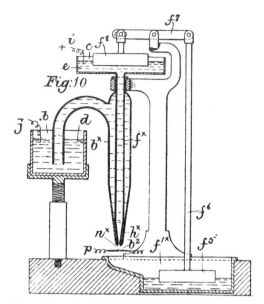

Fig. 6.17   *Alternative version of the 'Armorl' receiver, where the mercury and acid are contained in concentric tubes*
[British Patent 21 981, 1901]

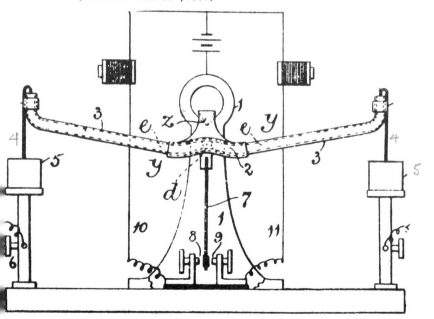

Fig. 6.18   *Balanced beam type of 'Armorl' detector*
[British Patent 21 981, 1901]

device. The mercury runs into the sump at the bottom, and as the level there rises the float (f^5) rises, and via the mechanical linkage depresses the plunger (f^8) which keeps the level constant in the upper chamber.

The version shown in Fig. 6.18 works in rather a different way. Here a shaped glass tube (3) is balanced on a knife-edged support (z). It contains a small bubble of acid (d) trapped between two columns of mercury (e) and normally it rests in a horizontal position. When a voltage is applied between the columns of mercury (contact being made by means of the wires (4) dipping into the mercury cups (5)) the surface tensions change so that the bubble of acid is displaced and the tube tilts and makes one of the contacts (8 and 9) at the bottom. The patent goes on to explain how the sensitivity of the apparatus may be improved by making the ends of the tubes thinner as shown in Fig. 6.19 so that a small amount of mercury displaced flows out to a greater distance from the pivot and thus creates more torque. In a second patent Orling suggested further modifications to increase the sensitivity even more.[44]

It must be stated again quite clearly at this point that these devices were relays intended for use with the Armorl system which was a d.c. earth-current signalling system. Two metal spikes were driven into the ground some distance apart at the transmitting station, and a direct voltage was applied to them.[45] Two similar spikes at the receiver would have a small voltage induced in them by the earth currents, and the relays were intended to record the presence or absence of this minute voltage. One account speaks of success up to a distance of 5 miles using a 10 yard base line, and refers rather caustically to 'the advances, if any, made by Armstrong and Orling'.[46] It is quite clear that the apparatus of Fig. 6.18 in particular is totally symmetrical in construction, so it could not possibly have worked with alternating currents. Indeed, in their patent they deal with oscillatory aerial currents by employing a normal filings coherer, and they suggest using the Armorl principle to produce the decohering shock. One suggestion is shown in Fig. 6.20. The vessel contains a layer of acid (d) lying above a layer of dense fluid (d') which, in turn, lies above a layer of mercury. The cylindrical tube (A) lying in the insulating layer is just an ordinary filings coherer tube. The idea here is that when the coherer is brought to the low-resistance condition by an oscillatory signal a direct voltage is applied to the capillary (f), which then allows a drop of mercury to fall. This hits the coherer and taps it back to the high-resistance state.

This is clear evidence that as far as the inventors were concerned, the Armorl devices were a means of detecting small direct voltages, and that when alternating voltages were involved they considered it necessary to

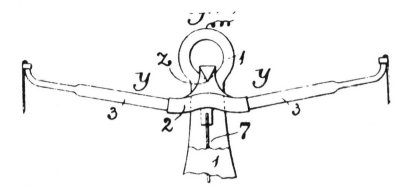

Fig. 6.19 *Improved tube for the detector of Fig. 6.18*
[British Patent 21 981, 1901]

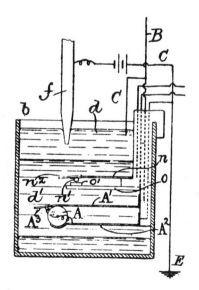

Fig. 6.20 *'Armorl' mercury drop principle used to produce decoherence in the filings coherer tube (A)*
[British Patent 21 981, 1901]

introduce a coherer tube. Once again, however, we have a curious confusion in contemporary accounts which suggests that it could be used with oscillatory voltages also. Mazzotto,[47] for example, stated that 'The Armorl electrocapillary relay serves as a wave detector and may be used instead of a coherer' (no mention of a crystal rectifier here). Bottone,[48] who translated Mazzotto's book into English, says 'so great is the sensitiveness of this relay that it may take the place of a coherer in wireless telegraphy'. A.T.Story,[49] in his little book entitled *The story of wireless telegraphy* says

> Up to the present time the Armorl system has not been worked at a distance of more than 20 miles. Beyond that distance the inventors are compelled to have recourse to relays (meaning relay stations presumably), or else avail themselves of aerial transmission. For this purpose they require a special installation with high poles etc. just like Marconi and others, but they claim as an advantage the fact that the poles are only one tenth the height of those of Marconi. They claim further advantages in the speed of signalling.

This account seems only to make confusion worse confounded.

It must simply be admitted that the Lippmann types of receivers (and their Armorl derivatives) are most unlikely to have been usable with unrectified a.c. signals, although we do have all these indirect hints of their having been so used. In the absence of other information or of further experiments it is probably safest to conclude that these writers misunderstood the information they had been given.

To summarise: although several of these thin-film and capillary devices appear to have worked fairly successfully, the only one which went into practical service to any extent seems to have been the Lodge–Muirhead–Robinson wheel detector of Fig. 6.3.

References

1   APPLEYARD, R.: *Phil. Mag*, **43**, 1897, p.374
2   APPLEYARD, R.: *Electrician*, **38**, 1897, p.800
3   *Science Abstracts,* **3**, 1900, no. 1464, p.545
4   CAMPANILE, F., and DI CIOMMO, G.: *Electrical Review* (New York), **36**, 1900, p.333
5   WILSON, E., and EVANS, R.J.: British Patent 30 846, 1897
6   LODGE, O.J.: *Proc. Roy. Soc.* (London), **71**, 1903, p.402
7   DE TUNZELMAN, G.W.: *Wireless telegraphy – a popular explanation* (Office of Knowledge, London, 1901), p.67
8   LODGE, O.J. *et al.* British Patent 13 521, 1902
9   POWELL, S.M.: *Electrical Review* (London), **68**, 1911, p.11
10 Reference 6

11 *Electrician,* **52**, 1903, p.85
12 MONKTON, C.C.F.: *Radio telegraphy* (Westminster Series, Constable, 1908), p.170
13 HOWGRAVE-GRAHAM, R.P.: *Wireless telegraphy for amateurs* (Percival Marshall, 1907 [?]), chap.4
14 ECCLES, W.H.: *Electrical Engineering,* **1**, 1907, p.241
15 BLAKE, G.G.: *History of radio telegraphy and telephony* (Chapman and Hall, 1928), p.94
16 FESSENDEN, R.A.: US Patent 1 042 778, 1912
17 ECCLES, W.H.: *Wireless telegraphy and telephony* (Benn, 1918), p.322
18 FESSENDEN, R.A.: British Patent 11 154, 1910
19 WALTER, L.H.: British Patent 17 111, 1902
20 WALTER, L.H.: *Electrician,* **52**, 1903, p.173
21 LODGE, O.J.: *JIEE,* **19**, 1890, p.346 (see also p.352)
22 LODGE, O.J.: *Electrician,* **40**, 1897, p.87
23 RAYLEIGH: *Phil. Mag.,* **48**, Ser. 5, 1899, pp. 321, 337
24 *Science Abstracts,* **3**, 1900, no. 13, p.5
25 RAYLEIGH: *Proc. Roy. Soc.* (London), **28**, 1879, p.406
26 RAYLEIGH: *Proc. Roy. Soc.* (London), **34**, 1882, p.130
27 BOYS, C.V.: British Patent 13 828, 1905
28 Reference 17, p.323
29 Reference 15, p.87
30 LIPPMANN, G.: *Journal de Physique Theorique et Applique,* **3**, 1874, p.41
31 LIPPMANN, G.: *Phil. Mag.,* **4**, Ser. 5, 1877, p.239
32 LIPPMANN, G.: *Ann. de Chimie et de Physique,* **12**, Ser. 5, 1877, p.265
33 BURCH, G.J.: *Electrician,* **37**, 1896, pp.380, 401
34 VANNI, G.: *Electrician,* **41**, 1898, p.579
35 ORLING, A., and ARMSTRONG, J.T.: British Patent 19 063, 1903
36 ROUSSEL, J.: *Wireless for the amateur* (Constable, 1923), p.256ff
37 Reference 15, p.20
38 *La Lumiére Electrique,* **16**, 1885, p.428
39 JENTSCH, O.: *Telegraphie und Telephonie ohne draht* (Springer, 1904), p.177
40 *Electrical Magazine,* **1**, 1904, p.176 (Membrane omitted in error from diagram in this reference)
41 ERSKINE-MURRAY, J.: *Wireless telegraphy*, (Crosby-Lockwood, 2nd edn., 1909), p.138 (4th edn., 1913), p.131
42 *Electrical Review* (London), **55**, 1904, p.283
43 ORLING, A., and ARMSTRONG, J.T.: British Patent 21 981, 1901
44 ORLING, A.: British Patent 10 327, 1906
45 STORY, A.T.: *The story of wireless telegraphy* (1st edn., Newnes, 2nd edn., Hodder and Stoughton, undated), chap. 12
46 *Electrical Review* (London), **49**, 1901, p.1061
47 MAZZOTTO, D.: *Wireless telegraphy and telephony* (Whittaker, 1906), p.205
48 BOTTONE, S.R.: *Wireless telegraphy and Hertzian waves* (Whittaker, 1910, 4th Edn.) p.129
49 Reference 45

Chapter 7

Thermal detectors

Most of the detectors which have been described so far were intended simply to indicate the presence or absence of signals in the receiving aerial. The thermal detectors which will be described in this Chapter were different in that they were also able to provide a quantitative measurement of the strength of the received signal. By and large, they were not as sensitive as some of the other detectors, or as widely used in practical signalling systems, but they were greatly valued for experimental purposes because of this almost unique property.

The principle on which they operated was very simple. A current flowing in a wire produces heat, and this heat may be measured in some way to reveal the presence of the current and to estimate its magnitude. Since the heating effect is proportional to the square of the current the instrument may be calibrated by use of a known direct current, and it will then read the root-mean-square value of the alternating current. There was nothing particularly new in this principle, of course; the hot-wire ammeter was already a well established instrument. The main problem for the radio experimenter was to adapt it to measure the minute currents and powers induced in a receiving aerial. Apart from one or two quite unworkable ideas such as applying the voltage directly to a small incandescent bulb and reading the Morse signals from its flashes,[1] the thermal detectors can be divided into several well-defined groups.

When a piece of metallic wire is heated its electrical resistance rises, and one of the simplest ways of keeping track of the temperature changes is to monitor this resistance. Instruments which operate in this way are often referred to as 'bolometers' or 'bolometric receivers', but

R.A.Fessenden thought this word inappropriate and specified in his patents that they should be called a 'barretters'. According to him, a bolometer is an instrument which is designed to receive radiant energy and to pass it on in another form (as an electric current, for example) with as great a conversion efficiency as possible. A barretter on the other hand is designed to retain as much of the energy as possible in order to achieve the maximum temperature change and hence the maximum resistance change.

There are several references in the early literature to the connection of a fine piece of wire across the gap of a simple Hertzian receiver or in an aerial/earth circuit.[2-5] Its resistance was measured by including it also in a Wheatstone bridge network. The occurrence of a pulse of radio-frequency signal would be indicated by sudden imbalance of the bridge, causing deflection of the galvanometer. This sounds very simple, but in practice two important factors have to be taken into account. When a current flows in a wire, the wire receives heat from that current, but it will also lose heat to its surroundings by conduction, radiation and convection. The final temperature to which it settles (and hence its resistance) will depend on all these mechanisms, and naturally draughts and ambient temperature changes will play an important part. If anything more than a simple yes/no indication is to be achieved something must be done to reduce these effects, or at least to keep them constant during the course of the measurements. The second factor is that, as mentioned above, the resistance change which occurs will depend on the temperature attained, and in the interests of sensitivity it is desirable to reduce the heat losses as far as possible.

With these points in mind Fessenden produced his barretter shown in Fig. 7.1.[6-11] The portion labelled (14) in the diagram is a very fine piece of wire made by a variant of the Wollaston process[12] already described in the Chapter dealing with electrolytic detectors. The silver-coated platinum wire was drawn down through a series of dies until it was as fine as possible. It was then attached to the stopper (19) and exposed to nitric acid fumes which left a very fine thread of platinum at the tip of the loop. This was said to have a diameter of 6×10^{-5} inches and a resistance of 30 Ω. It was enclosed in a silver thimble (18) which was in turn surrounded by an evacuated envelope (17) to keep the heat in and to isolate it from environmental changes. When in use it was connected in series with a resistor and battery (to provide a small direct current) and an earpiece; it was also connected between aerial and earth. The aerial current heated the loop and changed its resistance, and the resulting change in battery current was audible in the earpiece. This was obviously a very delicate piece of apparatus, and in the practi-

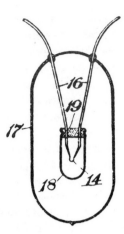

Fig. 7.1   *Fessenden's fine-wire barretter*
[British Patent 17 705, 1902]

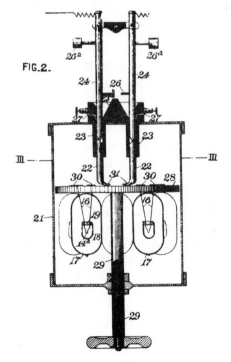

Fig. 7.2   *Several barretters mounted on a rotating turret for easy replacement
in the event of accidental destruction*
[British Patent 17 705, 1902]

cal system several barretters were arranged on a rotating turret as in Fig. 7.2 so that in the event of damage to the fine wire another one could be brought into service with the minimum delay.

A rather different approach to the question of ambient changes was used by several people[13-16] and is illustrated in Fig. 7.3. The rectangles R and S were made of very fine platinum wire and they formed two arms of a Wheatstone bridge. The resistors P and Q were adjusted to balance the bridge. When aerial current flowed down through R the change in resistance resulted in deflection of the galvanometer. The point of the second rectangle was, of course, that it would be subject to the same environmental changes as R so that spurious temperature variations would be cancelled out by a 'common mode' mechanism. A fact which is perhaps worthy of note in passing is that the rectangle itself forms a balanced bridge, and no radio-frequency current flows in the rest of the circuit. It might be said to be an early example of a balanced demodulator. This was quite a sensitive receiver, and Camille Tissot reported that he was able to receive signals from a distance of 50 km with a circuit of this sort.[17] To achieve good sensitivity, the rectangles had to be made of very fine wire and Fitzgerald wrote with some feeling, 'We intend using silvered quartz fibres, our hearts having been broken trying to use that brittle beauty Wollaston wire'.[18]

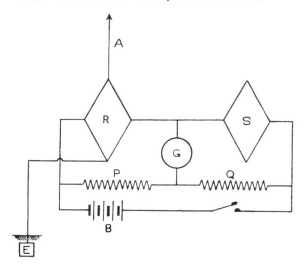

Fig. 7.3   *Rubens and Ritter thermal detector using a Wheatstone bridge circuit to detect resistance changes*
[Fleming, J.A.: *Principles of electric wave telegraphy and telephony* (Longmans, 3rd edn., 1916), p.514; Turpain, A.: *Les applications des ondes électriques* (Carre et Naud, 1908), p.33]

As mentioned previously, receivers of this sort using very fine wires were very delicate, and they were particularly susceptible to damage by atmospheric electricity. A good lightning flash in the vicinity would induce enough current in the aerial to vapourise the wire completely. To remedy this, Fessenden introduced his liquid barretters,[19-23] one of which is shown in Fig. 7.4. A glass vessel was divided into two parts by a glass plate and was filled with sulphuric acid. A small hole in the plate constrained any current flowing between the two platinum electrodes inserted at top and bottom to pass through a thin filament of liquid which behaved in the same way as the fine wire. Such a column of liquid was said to have a greater coefficient of change of resistance with temperature than a metal wire. Fessenden stated that a strong acid solution changed its resistance by 12% per $^\circ$C, compared with 0·5% per

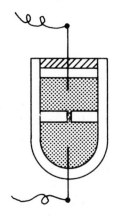

Fig. 7.4   *Liquid barretter*

$^\circ$C for a platinum wire. The supreme merit of this barretter was that in the event of destruction of the liquid filament by overload it was self-repairing, the liquid flowing back into the hole as soon as any vapour had disappeared. It will be observed that in this form the device looks very similar to the electrolytic detectors described in Chapter 4, and in fact several people, Fessenden included, believed that those detectors were thermal rather than electrolytic in action. As was explained in that earlier section, Ives and others proved conclusively that in the case of the fine-point detectors at least, the action was truly electrolytic, but one cannot rule out the possibility that in Pupin's electrolytic detector there might well have been some thermal action in addition to electrolytic effects which contributed to the operation.

Another thing which happens to a wire when it is heated is that it

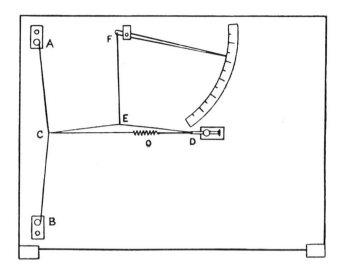

Fig. 7.5   *Fleming's hot-wire ammeter*
[Fleming, J.A.: *Principles of electric wave telegraphy and telephony* (Longmans, 3rd edn., 1916), p.229

expands in length, and this too can be used to measure the current as in the 'hot-wire' meter of Fig. 7.5. This particular device was an attempt by Fleming[24] to improve the sensitivity of the well-established hot-wire instrument to make it suitable for high-frequency detection and measurement. A loop of thread tensioned by the spring (*o*) passes over a fixed point (d) and around the wire or wires (A–B). From point E another thread is taken to the pointer which can move along a curved scale. Current flowing in the wire causes heating and expansion which in turn moves the pointer. Although the magnification provided by the thread arrangement and the leverage in the pointer might lead one to expect great sensitivity, from the account given by Fleming it does not seem to have been much use for currents less than about 0·1 A.

An alternative form of expansion meter was invented by W.G. Gregory[25] and is illustrated in Fig. 7.6. This made use of what was called a 'Perry' or 'Ayrton–Perry' magnifying spring. A long strip of

Fig.7.6   *Gregory's thermal detector using the Ayrton–Perry magnifying spring*
[*Proc. Phys. Soc.* (London), **10**, 1889, p.291]

metal foil is wound around a thin cylindrical object so that a helical spring is formed. This spring (S in the diagram) is fixed at one end, and to the other end is attached a long piece of platinum wire lying inside a glass tube, being fixed at the far end of the tube. In the particular version described in Gregory's paper the Perry spring was about 25 cm long and the wire was 192 cm long. Changes in length of the wire caused by the heating effects of currents flowing in it caused the spring to coil or uncoil, and this rotational movement was observed by means of a mirror (M) mounted at the end of the spring. According to the inventor it was possible to observe a change in length of 5×10^{-6} mm resulting from a temperature change of $0.003°C$, and by modifying the position of the centre of gravity of the mirror it was possible to increase the sensitivity further by a factor of about ten.

In spite of the fact that ambient temperature changes in the wire were compensated by expansion of the glass tube and the brass cylinder in which the Perry spring was mounted, it must have been rather difficult to keep the light beam reflected from the mirror on its scale zero. The screw arrangement at the left of Fig. 7.6 was provided as a zero adjustment. In spite of the apparent sensitivity, Gregory does not seem to have been all that successful in receiving radio signals with it as he reported reception (using the unmodified model) over a distance of only 4 m.

The drift problem was also in the mind of Duddell[26-29] who con-

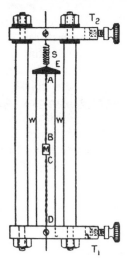

Fig. 7.7   *Duddell's double-spring detector*
        [*Proc. Phys. Soc.* (London), **19**, 1904, p.238]

structed the instrument shown in Fig. 7.7. The ends of the two wires
(W) were fixed at the top to a small bar of ebonite (E) and at the
bottom to the block of ebonite (T_1). Two Perry springs (AB and CD)
wound in opposite directions were also stretched between the block
and the bar, and the whole thing was kept under tension by the light
spring (S). A small mirror (M) was mounted at the centre of the springs.
With this arrangement, current passing through the wires and altering
their lengths causes rotation of the mirror. On the other hand, since the
Perry springs and the wires are made of the same material, room temperature changes common to both cancel out and result in no rotation.
According to Duddell, this could be used to detect currents as low as
0·5 mA, or powers down to 400 μW. The instrument itself is shown in
Fig. 7.8. The mechanism was protected by covers at front and back,
and it seems to have been quite a robust affair because the inventor
claimed to have carried it about in his pocket for three years without its
being damaged or suffering any ill effects.

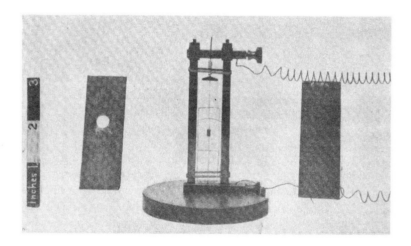

Fig. 7.8   *Duddell's double twisted-strip thermal detector with front and back
covers removed to show construction*
[Wireless World, 5, 1917/18]

Another type of detector which seems to have been invented independently by Professors Threlfall and Fleming is shown in Fig. 7.9.[30-34]
Two horizontally mounted wires support a small flat mirror, and the
whole assembly is kept taut by means of a small weight or spring. In
the event of an ambient change in temperature causing equal sag in
both wires, the mirror will still point in the same direction, but if an

aerial current is passed through one wire only, differential sag occurs and the resulting rotation of the mirror may be observed with the aid of a light-beam in the traditional manner. An alternative method of ob-

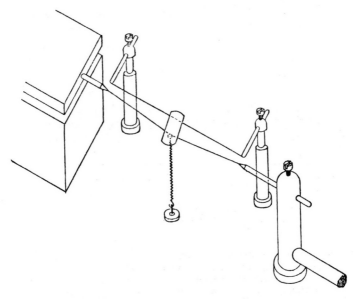

Fig. 7.9   *Hot-wire ammeter constructed by Threlfall*
[*Proc. Phys. Soc.* (London), **19**, 1904, p.59]

servation is to provide the active wire with a micrometer adjustment at one end which can be used to tighten the wire and to restore the deflected light spot to its former position. The micrometer can be calibrated to read the value of the current directly. Fleming claimed that his instrument was able to measure radio-frequency currents down to about 2 mA; Threlfall claimed that his was usable for potential differences as low as 0·03 V. It is, perhaps, only fair to stress here that these instruments, together with that of Duddell (previously described) were conceived primarily as general methods of measuring small high-frequency currents. Their use in radio reception and measurement was simply one particular application.

A slight variation on the theme of the expanding wire was the hot-wire telephone invented by Preece,[35-37] and shown in Fig. 7.10. This was intended originally as a telephone receiver for audio-frequency signals, but it was used by Blake and by Campbell-Swinton for direct detection of radio signals. The thin wire (P) is fixed firmly at one end and the other end is fastened to the centre of a diaphragm which moves when the wire changes length. With such a receiver connected between aerial and earth Morse signals were, apparently, clearly audible.

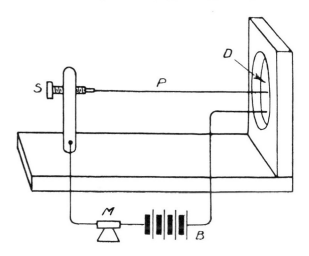

Fig. 7.10 *Preece's thermal telephone receiver*
Changes in length of wire P were audible at diaphragm D
[Blake, G.G.: *History of radio telegraphy and telephony* (Chapman & Hall, 1928), p.21]

If a heated wire is contained in a closed vessel filled with air then the air will be warmed and its pressure will increase. This change in pressure can be measured with a manometer, as shown in Fig. 7.11. This arrangement was originally devised by W.Snow-Harris, Surgeon, of Plymouth in 1827 as an aid to his investigations into 'the relative powers of various metallic substances as conductors of electricity'. It was used by various people as a radio detector and appeared in a number of forms; the one shown is Snow-Harris's original.[38-41] The German P.T. Riess also described a very similar piece of apparatus in his book published in 1853,[42] and in the German literature it is usually referred to as the Riess thermogalvanometer. Changes in room temperature were obviously a nuisance and caused variations in the position of the liquid column. Riess was responsible for the introduction of the differential model shown in Fig. 7.12.[43] This had two air vessels (*a* and *b*) working in a balanced manner so that spurious temperature changes were cancelled out.

Expansion of the air was also the principle which lay behind the two receivers of Figs. 7.13 and 7.14.[44-48] The De Lange thermal telephone

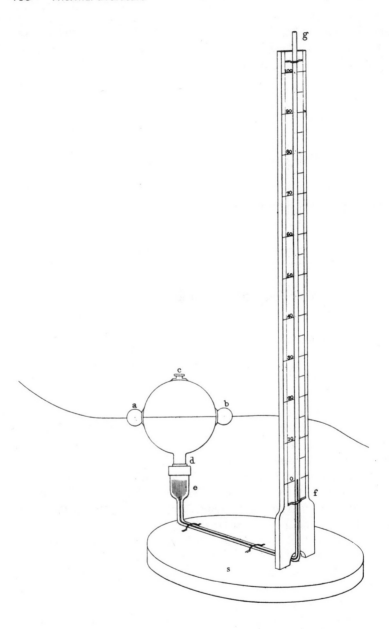

Fig. 7.11   *Thermogalvanometer made in 1827 by Snow-Harris to aid investigation into ' the relative powers of various metallic substances as conductors of electricity'*
[*Phil. Trans. Roy. Soc.,* **117**, 1827, p.18]

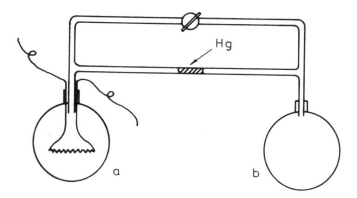

Fig. 7.12   *(Riess's balanced thermal galvanometer to counteract the drifts due to changes in ambient temperature*

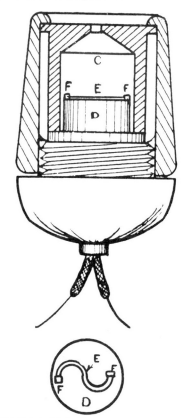

Fig. 7.13   *De Lange thermal telephone*
[Coursey, P.R.: *Telephony without wires* (Wireless Press, 1919), p.330]

was made out of solid marble with an ebonite screw-on cap. It contained a fine wire (E) through which the aerial currents passed and was designed to be held to the ear so that the expansion and contraction of the air (and hence the Morse signals) could be heard directly. The Eccles Thermophone was similar in principle, but the wire was contained in a glass tube of such dimensions that it could be inserted directly into the ear. In order to produce an adequate sound both these receivers had to contain a very fine wire as the active element, and this meant that they were very delicate and suffered continual destruction by atmospherics, so that although they undoubtedly worked, they were little used in practical signalling systems.

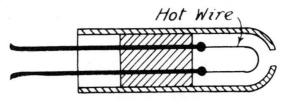

Fig. 7.14   *Eccles thermophone*
It was sufficiently small to enable it to be inserted into the ear
[Blake, G.G.: *History of radio telegraphy and telephony* (Chapman & Hall, 1928), p.21]

One other well known fact about hot air is that it rises. This was exploited in the very sensitive detection apparatus devised by C.V. Boys and his colleagues in 1890.[49] This was a modification of an instrument originally invented by Joule. The Joule instrument is said to have been so sensitive that it was able 'to detect the direct heating of the moon's rays unconcentrated by lens or mirror'. A vertical cylindrical pipe made of cardboard is closed at the top, as shown in Fig. 7.15*a* and is divided into two portions by a central partition (see plan view above). This partition does not quite reach to the top and there is a small gap through which the air can move. Immediately below the partition there is a small phial containing phosphorus. One end of this is open to the air so that the phosphorus oxidises gradually and a thin stream of white oxide vapour falls vertically as shown. A fine platinum wire is situated in one of the chambers. If this is heated by a current, natural circulation occurs as the warm air rises and this deflects the oxide stream from the vertical. The extent of the deflection can be read off the scale, and this can be calibrated beforehand with a direct current if required. According to the inventors, a deflection of 1 mm was caused by a power of 0·000 025 calories per second, which, in modern terms, is equivalent to 0·1 mW. One of the main disadvantages with this arrangement is that the phos-

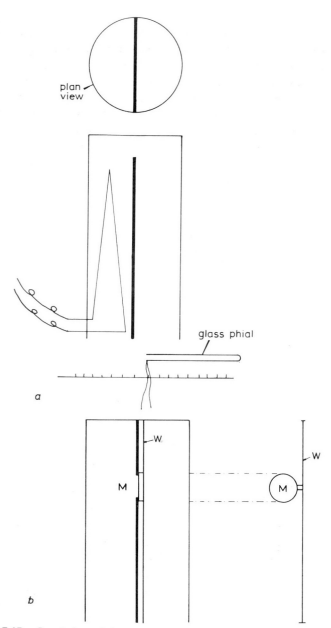

plan
view

glass phial

a

W

M

W

M

b

Fig. 7.15   *Boys's thermal detector*
(a)   Initial version in which air movement deflects the stream of
        smoke issuing from the glass phial
(b)   Modified version in which air movement deflects the trapdoor
        (M) suspended on the torsion wire (W)

phorus will only produce the stream of oxide if the temperature is greater than about 15°C, so that it is only usable on warm days. In an alternative version which is illustrated in Fig. 7.15*b* the division in the cylinder is complete, but air flow between the chambers is still possible via a small hole in the partition. A light mirror hanging on a torsion suspension wire W acts as a trapdoor across this hole. Air movement causes the door to swing open, the extent of the movement depending on the degree of heating. This version is reported as having been able to detect a power dissipation as low as 0·000 014 calories per second, which is just over one calorie per day (in modern terms. 0·06 mW).

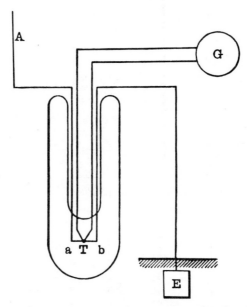

Fig. 7.16   *Hot wire/thermocouple detector constructed by Fleming*
[Fleming, J.A.: *Principles of electric wave telegraphy and telephony* (Longmans, 3rd edn., 1916), p.518]

Apparatus of this sort is bound to be extremely sensitive to general turbulence in the air, and elaborate precautions had to be taken to obviate errors from this source. The whole was protected by an outer glass cylinder. This was made to rotate slowly (1 revolution in 8 seconds) to ensure that it was at an even temperature and to eliminate the possibility of one side being warmer than the other due to heat or light radiations falling on it. This would set up internal air currents and disturbed the oxide stream or the trapdoor.

A rather more convenient way of measuring the rise in temperature for a wire is to use a thermocouple, a method which seems to have been

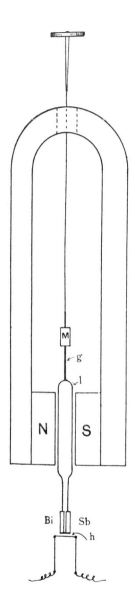

Fig. 7.17   *Duddell's modification of the Boys radio micrometer*
[Proc. Phys. Soc. (London), **19**, *1904, p.241]*

first introduced by Klemencic in 1891, and in an improved form by Fleming in 1906.[50-52] In Fleming's version, which is shown in Fig. 7.16, the oscillatory current was passed through a thin resistive wire of Constantan (a copper–nickel alloy) to the centre of which was fastened a bismuth/tellurium thermocouple, the whole being contained *in vacuo*. A low-resistance galvanometer connected to the thermocouple provided an indication of the wire temperature and could be calibrated to read the current in the wire.

A very sensitive thermocouple instrument was constructed by Duddell,[53-54] and in a slightly different form by Pierce.[55] It was a modified version of an instrument known as Boys radiomicrometer, and like some of those previously mentioned, it was first conceived as a general h.f. measuring instrument. It is shown in Fig. 7.17 and 7.18. As many a schoolboy has learnt from his physics lessons, the Boys original instrument was sufficiently sensitive to detect the heat of a candle at a distance of nearly two miles, and in a more refined form, at a distance

Fig. 7.18   *Duddel/Boys thermogalvanometer (with cover)*
[Wireless World, 5, 1917/18]

of 15 miles.[56] In the Duddell version an antimony/bismuth (Sb/Bi) thermocouple is connected to a loop of wire which is suspended from a torsion fibre in a magnetic field. The junction of the thermocouple is situated just above a resistive wire which, in this case, is a strip of gold foil. Heat generated in the foil causes current to flow in the loop which then rotates, its movement being observed by a mirror (M) and light beam in the usual way. This instrument was said to be able to measure powers dissipated in the foil as low as 0·5 μW.

Fig. 7.19 shows another thermocouple arrangement attributed to Douda and to Brandes.[57-58] A thin wire (*ds*) made of copper or iron is bent to form a right angle. A Constantan wire (*d's'*), also bent into a right angle, is looped under it, and the two wires are kept in tight contact by means of a spring. An oscillatory current flowing from *d* to *d'*

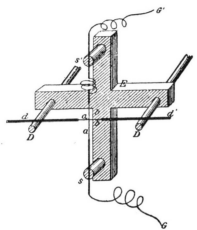

Fig. 7.19 *Thermocouple detector attributed to Brandes and to Douda* [de Valbreuze, R.: Notions générals sur la t.s.f. (L'Eclairage Electrique Paris, 1907), p.101]

warms up the wire by the usual Joule i^2R heating; this heats the copper/Constantan junction, causing an e.m.f. to appear between *s* and *s'*.

Since a thermocouple is itself made out of two pieces of wire it is possible to change the arrangement of Fig. 7.20*a* to that of Fig. 7.20*b*. Here the resistance wire has been done away with altogether and the thermocouple acts as its own resistance wire. Current passing through it will cause normal Joule heating. The heat produced will tend to leak away from the thermocouple along the connecting wires so that a thermal gradient is established. The junction at the centre will then be hotter than the ends so that a thermal e.m.f. is produced which is observed by the galvanometer connected as shown. (It is also possible for a junction of this sort to be cooled or heated by the Peltier effect, depending on the direction of current flow. In this case where alternating current from the aerial flows, the heating and cooling cancel out; in any case, the Peltier effect is much smaller than the i^2R heating effect).

As an example of a junction used in this way, Professor Austin[59-61] used a thermocouple consisting of a piece of tellurium metal pressed firmly against the rim of a rotating aluminium disc or a pointed aluminium rod, the rotation serving to ensure a good metal-to-metal contact at all times. Other popular combinations[62–65] used the junction between a metal and silicon, or between a metal and the crystalline substance chalcopyrites (a mixed sulphide of iron and copper).

Readers who are familiar with 'cat's whiskers' will now appreciate that we have strayed into a very gray area of the subject. A junction between two different metals, or between a metal and crystalline substance, can exhibit the property of rectification, and it is but a short step to convert the diagram of Fig. 7.20*b* to that of 7.20*c*. This being so, it is now pertinent to ask whether the deflection of the galvanometer in 7.20*b* is really due to a thermojunction effect as described, or whether it is caused by the presence of a rectifying diode. The problem is summed up by a quotation from a paper written by S.M. Powell in 1911.[66] He says

> Thermoelectric detectors *must* be carefully distinguished from the unilateral or 'rectifying' types, although the types are practically identical in construction; and in many cases the detector action is due to a combination of thermal and rectifying effects, while in a number of cases it is by no means certain whether the action should be considered thermal or rectifying.

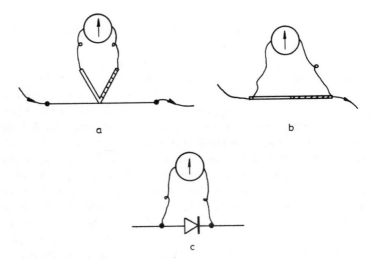

Fig. 7.20   *Transition from hot-wire and thermocouple to the semiconductor diode*
(*a*)   Thermocouple attached to hot wire
(*b*)   Thermocouple forms its own hot-wire
(*c*)   Thermocouple junction replaced by diode

This 'broadly speaking, yes and no' statement may appear to be none too helpful! However, it does summarise the position quite well. In many instances where junctions were used in this way both effects were undoubtedly present, and it was a matter of degree as to which one predominated in any given case. Actually, some writers at that time believed anyway that the phenomenon of rectification could be explained on purely thermal grounds. W.H.Eccles[67] was still suggesting this in a book published as late as 1933. Eccles seems to have had a great liking for thermal explanations of phenomena found in the various types of detector, and he believed that electrolytic detectors could be explained in this way as we have already mentioned. To be fair to him, in his 1933 book he did go on to say 'but a more fundamental explanation (of crystal rectification) is awaited, perhaps based on the modern theory of steep potential gradients at the boundaries of conductors' a suggestion which begins to sound very much like the semiconductor theory of today.

Thus we may sum up the properties of the thermal detectors of Hertzian waves as follows: Their most useful property was that they could provide a quantitative measure of the strength of the received signal and for this reason they were widely used in experimental and laboratory work (and thermocouple instruments in particular are still so used). However, they suffered from several disadvantages. They tended to be inconvenient to use because of all the elaborate precautions which had to be taken if accuracy and repeatability of results were to be achieved. Because of the need to use fine wires for good sensitivity they were very delicate and could easily be damaged by mechanical and electrical shocks. Although some of them were very sensitive, taking the group as a whole they were not as sensitive as other detectors such as the coherers or the electrolytic detectors.

Finally, one factor which we have not yet mentioned: many of them were very slow acting. As an example of this we may take the fact that the Duddell thermocouple instrument of Fig. 7.17 required ten seconds to settle to its final reading. The twisted-strip instrument of Fig. 7.7 was better, but even this had a time constant of about 1/15th of a second, so that it was not able to respond to signals switching on and off at rates greater than about twice per second. This slow response made the thermal instruments unsuitable for signalling purposes, and with the possible exception of Fessenden's barretters, they were very little used outside the laboratory.

References

1   TOWNSEND, J.S.E.: British Patent 130 429, 1918
2   FITZGERALD, G.F.: *Nature*, **41**, 1890, p.295
3   PRESTON, T.: *Theory of light* (Macmillan, 3rd edn., 1901), p.552
4   *Electrical World and Engineer*, **36**, 1900, p.387
5   TAYLOR, A.H.: *Phys. Rev.*, **18**, 1904, p.230
6   FESSENDEN, R.A.: US Patent 706 744 1902
7   FESSENDEN, R.A.: US Patent 706 742, 1902
8   FESSENDEN, R.A.: British Patent 17 705, 1902
9   MONKTON, C.C.F.: *Radio telegraphy* (Constable, 1908), p.176
10 SEWALL, C.H.: *Wireless telegraphy* (Crosby–Lockwood, 1904), p.167
11 COLLINS, A.F.: *Electrician*, **49**, 1902, p.945
12 WOLLASTON, W.H.: *Phil. Trans. Roy. Soc.* (London), **103**, 1813, Pt. I, p.114
13 TISSOT, C.: *Manuel élémentaire de la télégraphie sans fil* (Challamel, Paris, 5th edn., 1918), p.113
14 STONE, J.S.: US Patent 767 981, 1904
15 TISSOT, C.: *JIEE*, **36**, 1905/6, p.448
16 RUBENS, H., and RITTER, R.: *Wied. Ann.*, **40**, 1890, p.56
17 FLEMING, J.A.: *Principles of electric wave telegraphy and telephony* (Longmans, Green, 3rd edn., 1916), pp.237 and 514
18 Reference 2
19 FESSENDEN, R.A.: US Patent 12 115 Reissued, 26th May 1903
20 FESSENDEN, R.A.: US Patent 731 029, 1903
21 FESSENDEN, R.A.: British Patent 28 291, 1903
22 WALTER, L.H.: *Electrical Magazine*, **2**, 1904, pp.384 and 596
23 ZENNECK, J., and SEELIG, A.E.: *Wireless telegraphy* (McGraw-Hill, 1915), p.281
24 FLEMING, J.A.: *An elementary manual of radio telegraphy and telephony* (Longmans, Green, 1911), p.274 (or Reference 17, p.229)
25 GREGORY, W.G.: *Proc. Phys. Soc.* (London), **10**, 1899, p.290
26 BOUASSE, H.: *Oscillations électriques* (Delagrave, Paris, 1924), p.273
27 DUDDELL, W.: *Proc. Phys. Soc.* (London), **19**, 1904, p.233
28 DUDDELL, W.: *Phil. Mag.*, **8**, Ser. 6, 1904, p.91
29 MARCHANT, E.W.: *Wireless World*, **5**, 1917/18, p.672
30 THRELFALL, R.: *Proc. Phys. Soc.* (London), **19**, 1904, p.58
31 THRELFALL, R.: *Phil. Mag.*, **7**, Ser. 6, 1904, p.371
32 FLEMING, J.A.: *Phil. Mag.*, **7**, Ser. 6, 1904, p.595
33 Reference 26, p.272
34 Reference 17, p.229
35 PREECE, W.H.: *Proc. Roy. Soc.* (London), **30**, 1880, p.409
36 BLAKE, G.G.: *History of radio telegraphy and telephony* (Chapman and Hall, 1928), p.92
37 BLAKE, G.G.: *The model engineer and electrician*, **31**, 1914, p.52
38 SNOW-HARRIS, W.: *Phil. Trans. Roy. Soc.* (London), **117**, 1827, p.18
39 BOULANGER, J., and FERRIÉ, G.: *La télégraphie sans fil* (Berger–Levrault, Paris, 1907), p.83

40   Reference 17, p.234
41   ERSKINE-MURRAY, J.: *Handbook of wireless telegraphy* (Crosby, Lockwood, 2nd edn., 1909), p.127
42   RIESS, P.T.: *Die lehrer von der Reibungselectricität* (Hirschwald, Berlin, 1853), Vol. 1, Fig. 97
43   Reference 26, p.270
44   COURSEY, P.R.: *Telephony without wires* (Wireless Press, 1919), p.330
45   *Electrician,* **74**, 1914, pp.358, 362 and 401
46   British Patent 3 954, 1915
47   ECCLES, W.H.: *Handbook of wireless telegraphy* (Benn, 2nd edn., 1918), p.322 and 356
48   Reference 36, pp.21 and 110
49   BOYS, C.V. *et al.: Proc. Phys. Soc.* (London), **11**, 1890, p.20
50   KLEMENCIC, J.: *Wied. Ann.,* **42**, 1891, p.417
51   Reference 24, p.199
52   Reference 17, p.517
53   References 27–29
54   SHOEMAKER, H.: *The Marconigraph,* **1**, 1913, p.556
55   *Electrician,* **45**, 1900, p.79
56   *Science Abstracts,* **2**, 1899, no. 1841, p.817
57   DE VALBREUZE, R.: *Notions génénerales sur la télégraphie sans fil* (*L 'Éclairage Électrique,* Paris, 1907), p.100
58   Reference 26, p.279
59   Reference 9, p.177
60   Reference 24, p.201
61   AUSTIN, L.W.: *Phys. Rev.,* **24**, 1907, p.508
62   PICKARD, G.W.: *Electrical World,* **48**, 1906, p.1003
63   ECCLES, W.H.: *Electrician,* **60**, 1908, p.587
64   AUSTIN, L.W.: British Patent 4 338, 1907
65   PICKARD, G.W.: British Patent 18 842, 1907
66   POWELL, S.M.: *Electrical Review,* **68**, 1911, pp.11 and 72
67   ECCLES, W.H.: *Wireless* (Butterworth, 1933), p.85

Chapter 8

Tickers, tone-wheels
and heterodynes

It was explained earlier that most of the early detectors were designed to produce a visible movement or a click in an earphone each time a burst of oscillations appeared in the receiving aerial. The dots and dashes of a Morse signal were represented by short or long trains of such bursts and were heard as noises which were more or less musical depending on the regularity with which the bursts were generated at the transmitter. However, when the art of transmitter construction had progressed to such a stage that it was possible to generate a continuous carrier wave, the reception of Morse signals became rather more difficult. Most of the existing detectors would produce a click at the beginning of each dot or dash as the carrier was switched on, to be followed by a period of silence until the end of the symbol when another click signified that it had been switched off again. One method of overcoming this difficulty was to chop the carrier wave on and off at the transmitter at audio rate during the course of the symbol. In modern terminology, the carrier itself was amplitude modulated with an audio-frequency square wave before the Morse was keyed on to it. This meant that once again a series of clicks would be heard at the receiver during reception of the individual symbol. The chopping process could be carried out at the receiver instead of the transmitter, and as late as 1920. A.N.Goldsmith[1] and the Marconi Co. were granted a patent for a means of doing this whereby a toothed wheel made of magnetic material was made to rotate between the primary and secondary coils of the aerial transformer. When a tooth intervened, it formed a magnetic screen and there was no coupling between the coils, but when a gap was present normal transformer action took place.

A more common method of dealing with a continuous carrier was to use a 'ticker', this word being an abbreviation of the German word 'tickerschaltung' – a ticking contact.[2] (It was sometimes written 'tikker' in the literature.) This was a device for which several patents were granted,[3-7] the earliest being one with a rather vague specification granted to Tesla in 1901. The principle of operation of a ticker is illustrated in Fig. 8.1. Oscillations from the aerial set up currents in the mutually-coupled resonant secondary circuit. This tuned circuit is periodically connected to a capacitor (C) by the vibrating contacts (K) which open and close at audio frequency. When the contacts are closed, the capacitor behaves as part of the tuned circuit, but when they open, a certain amount of charge is left on C depending on the state of oscillation in the circuit at the instant of opening. This charge leaks away through the earphone during the period of disconnection. (In practice, the frequency of oscillation would be very much higher relative to the contact opening frequency than can be shown in a diagram such as Fig. 8.1). If it were possible to arrange for the contact to open at *exactly* the same point on the cycle of the sinewave each time, then the capacitor would be topped back up to the same voltage each time as in Fig. 8.1*a*. In practice the contacts would be arranged to open and close at slightly different points on the cycle each time so that the amplitude of the current through the earphone would vary slowly, and a 'beat' note would be heard. In fact, to express it in more familiar terms, the frequency of the note produced will be the difference between the carrier frequency and some harmonic of the contact vibration frequency.

This is a very interesting principle; unlike some other detectors such as the coherer where the incoming signal releases energy from a local cell, energy from the incoming wave itself is converted directly into audible form. The earphone is a very sensitive instrument – or rather the ear with which it is used is very sensitive – so that relatively small amounts of energy are necessary to produce an audible sound. A.A.Campbell-Swinton,[8] commenting on the sensitivity of the telephone/ear combination claimed that it was possible to hear powers as low as one billionth of a watt, and to make his point even more plainly he pointed out that this amount of power bears the same relationship to one watt as does one second to 3,200 years! Rupert Stanley[9] in his *Textbook of wireless telegraphy* stated that signals in ticker circuits were audible with powers in the aerial circuit as low as 4×10^{-10} W while E.W.Stone[10] reported that reception could be achieved at greater distances with tickers than with other detectors. This is perfectly credible when one remembers that other detectors often have thresholds, and signals below the threshold cannot release energy from the local battery so that reception fails.

An additional advantage of the ticker method was that it was possible to distinguish between wanted and unwanted carriers of slightly different frequencies, as they would produce sounds of different pitches in the earphones.

There was one further advantage of the ticker which was stressed in the various patents which were granted. If a detector such as a coherer or an electrolytic cell is connected directly, or mutually coupled, to a tuned circuit its resistance damps that circuit making it less selective. This is particularly undesirable if the detector has a very variable and ill-defined resistance, as was the case with the coherer, because the tuning,

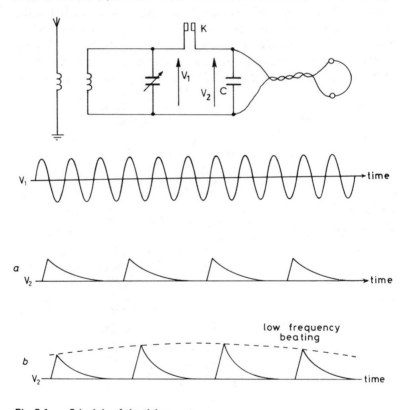

Fig. 8.1   *Principle of the ticker*
V_1 is a continuous sinusoid appearing across the tuned circuit
(a)   Capacitor voltage if contacts K close each time at the same point on the sinusoid
(b)   Capacitor voltage if contacts K close at different points on the sinusoid; a low-frequency 'beat' note is heard in the earphones

and hence the performance of the receiver, would vary with the random variations of the resistance. With the ticker method the tuned circuit can be isolated for a large part of the time so that it is able to build up resonant oscillations of large magnitude. The contacts are only closed briefly, so that the earphone is only connected for a small fraction of the time and reception benefits from this relatively undisturbed resonance, the capacitor (C) receiving a larger charge in consequence. The arrangement obviously makes for improved sensitivity and selectivity, and several of the patents suggested the use of ticker contacts with coherers etc. as well as simply with an earphone as in Fig. 8.1.

Having now described the principle of the ticker it is of some interest to look at the means which were used to effect the opening and closing of the contacts. The most obvious way of doing it was to attach them to a bell-trembler mechanism, as shown in Figs. 8.2 and 8.3.[11-14] A variant of this method is shown in Fig. 8.4;[15-16] here the left-hand contact behaves as a normal electric bell or buzzer. The armature at the right is also attracted and vibrated by the changing magnetism in the

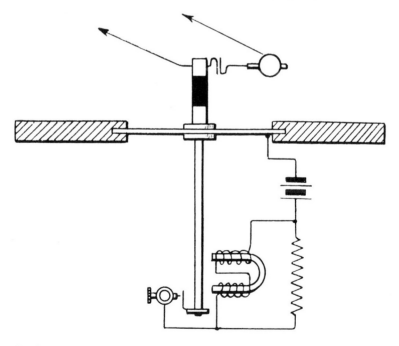

Fig. 8.2   *Ticker using a bell-trembler mechanism to operate the contacts (Poulsen)*
[Blake, G.G.: *History of radio telegraphy and telephony* (Chapman & Hall, 1928), p.96]

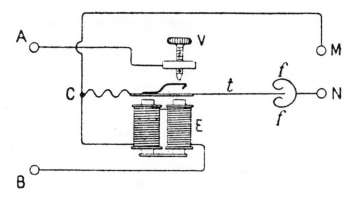

Fig. 8.3   *Bell-trembler ticker with double contacts (f) to double the speed of switching*
[Tissot, G.: *Manuel élémentaire de t.s.f.* (Chalamel, 1918), p.159]

iron core and its contacts are used in the ticker circuit. The reader will note the use of double contacts here and in Fig. 8.3 which serve to double the effective interruption rate. One experimenter[17] reported that he simply energised the coil by connecting it to the mains supply thereby doing away with the trembler contact and all the problems of adjustment and wear to which these are subject. This must have been a rather inflexible method of working as it allowed no opportunity of adjusting the interruption rate to suit the frequency of the carrier being received.

A very popular arrangement was the rotary form invented by Austin[18-21] shown in Fig. 8.5. Here a metal wheel is rotated by a small motor, and a fairly stiff piece of wire is clamped firmly at one end, the other being bent so that it rests in a small groove cut in the rim of the wheel. The direction of rotation is such that the wire 'chatters' against the wheel as it alternately catches and slips. The metal contact between the wire and the wheel makes and breaks in a manner ideal for

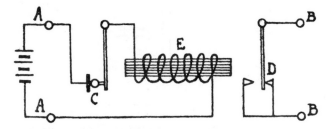

Fig. 8.4   *Alternative form of ticker using bell-trembler*
[Adam, M.: *Cours de t.s.f.* (Radio Home, Paris, undated), p.132

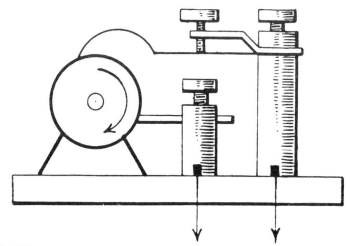

Fig. 8.5   *Austin's rotary ticker, with chattering spring contact*
[Stone, E.W.: *Elements of radio communications* (V. Nostrand, New York, 1926), p.531]

use in a ticker circuit. Actual photgraphs of a vibrator ticker and a rotary ticker may be seen in Fig. 8.6.

A slightly more exotic arrangement[22] is shown in Fig. 8.7. A glass tube placed above a gas flame behaves as a resonating organ pipe and emits a musical note. The gases in the pipe are highly ionised and two electrical connections as shown exhibit between them a resistance which changes in a periodic manner at the same frequency as the note emitted by the pipe. This, too, was used as a contact breaker for a ticker receiver, but with what success is not recorded.

A very common method was to use a toothed wheel rotating against a spring contact as shown in Fig. 8.8,[23 – 24] or the better-engineered

Fig. 8.6   *Two tickers used in the Poulsen system*
   (*a*)   *Vibrating ticker*
   (*b*)   *Chattering-contact ticker*

   [Wireless World V7, 1919-20, p.12]

version of Fig. 8.9, where the wheel consists of a thin cylinder having slots cut into its edge, the slots being filled with some insulating material and having a general appearance rather like the commutator of a direct-current motor.[25]

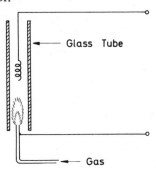

Fig. 8.7   *Horton's gas-flame ticker*

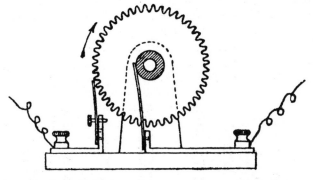

Fig. 8.8   *Toothed-wheel ticker*
[Adam, M.: *Cours de t.s.f.* (Radio Home, Paris, undated), p.132]

Most users of ticker receivers said that the sounds heard in the earphones were not nice musical tones as might have been expected from the explanation given in connection with Fig. 8.1, but were in fact buzzing or hissing noises which were not particularly pleasing to the ear.[26-27] This imperfection arose from the fact that bell tremblers and other devices of that sort do not have an exactly steady frequency of vibration. The frequency tends to vary around its average value in a random way. As has been pointed out, the audible sound has a pitch which is a beat between the carrier and a high harmonic of the contact frequency. A small variation of buzzer frequency can make a big difference to the note heard. Suppose, for example, that the carrier has a frequency of 30 000 Hz and that the contacts are vibrating at 200 Hz (a fairly typical rate). The beating will then be between the carrier and the 150th harmonic, which will give a beat of exactly zero frequency. Let the buzzer frequency now change by 1% to 202 Hz. The 150th harmonic of this will be 30 300 Hz so that the beat frequency will have changed from zero to 300 Hz. Other harmonics will also beat with the carrier to produce audible tones (the 151st for example) so that the net result will be a number of different notes each changing frequency as the buzzer frequency changes and merging to produce a rather un-musical buzz. It must also be remembered that mechanical contacts never open and close in a clean and precise manner but always have a certain amount of contact bounce, and this too would have added to the uncertainty of the final pitch.

The Goldschmidt tone-wheel was an attempt to improve matters in this respect.[28-31] It was very much like the ticker of Fig. 8.8 in principle except that it consisted of a commutator with a very large number of segments — 800 in the commercially produced model. It was rotated at a high speed, very carefully controlled and maintained, the general idea being that near synchronism with the carrier frequency could be produced. If it happened to rotate at exactly the right speed in synchronism with the carrier it could be arranged that only the positive parts of the carrier cycles would appear in the output so that halfwave rectification would result. If the speed were to be altered slightly from

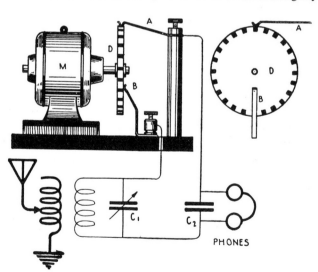

Fig. 8.9   *Poulsen's form of toothed-wheel ticker*
[Bucher, E.: *Practical wireless telegraphy* (Wireless Press, New York, 2nd edn., 1918), Section 221]

this state of synchronism, beating would occur and this would be heard in a receiver earpiece connected to the commutator brush. Since the speed of rotation was governed very closely and, in any case minor variations would be smoothed out by the flywheel effect of the rotating masses, the interruption rate was much more constant than in the simple tickers, and it was widely reported that the Goldschmidt tone-wheel produced a very acceptable musical sound. The apparatus itself is shown in Fig. 8.10.

Fig. 8.10   *Goldschmidt tone-wheel*
[Fleming, J.A.: *Principles of electric wave telegraphy and telephony*
[Longmans, 1916 p.705]

Among the advantages of the tone-wheel method of reception was that it was possible really to achieve the separation of wanted and unwanted signals by virtue of the different notes produced. This was not always possible with the simple tickers because of the imperfections already discussed. It was also possible to distinguish the musical tone produced by the signal from the noises produced by atmospheric disturbances. Furthermore, since every telephone receiver has its own particular resonance frequency at which it is most sensitive, it was now possible to adjust the pitch very carefully so as to obtain maximum audible effect. Working near synchronous speed also made better use of the energy in the received signal. In the case of a simple ticker using a low speed of interruption only a small fraction of the energy is transferred to the earpiece. At synchronous speed, the half wave rectification passes approximately half the energy to the earpiece. Even when the speed is altered slightly to produce an audible beat, high energy transfer is still obtained.

Even with a multisegment commutator it was necessary to rotate the shaft at very high speed in order to achieve near synchronism with a radio-frequency carrier. If, for example, a commutator with 800 segments were rotated at a speed of 3750 revolutions per minute (i.e. 62·5 revolutions per second) then it would be capable of detecting most efficiently a carrier at a frequency of about 50 000 Hz. Several suggestions were made as to how this high shaft speed could be reduced, one being that the frequency should be reduced in stages by passing the signal successively through three or more tone wheels connected in cascade.

An even more subtle method of using a lower speed of rotation was described by W.H.Eccles[32] (Fig. 8.11). Aerial and earth connections are taken to the slip-ring contacts S_1 and S_2 which connect with alternate segments of the rotating commutator. Let the frequency of an incoming sinusoid be ω_c. If the frequency of reversal between brushes B_1 and B_2 is at some nearby frequency $\omega_c + \delta$ then the interruptions produce, by multiplicative mixing, the frequencies $2\omega_c + \delta$ and δ, and if δ happens to be in the audio range it would be heard in a tele-

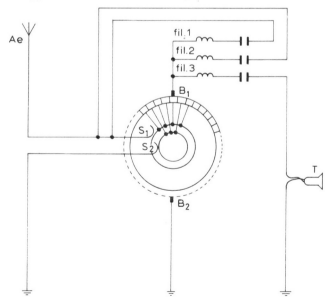

Fig. 8.11 *Multiple pass tone-wheel*
Brushes B_1 and B_2 bear on the segments of the rotating commutator

phone receiver connected across B_1 and B_2. Now let the shaft speed of the commutator be reduced so that the interruption is at some much lower frequency ω_i. This gives rise to sum and difference tones $\omega_c + \omega_i$ and $\omega_c - \omega_i$. The lower of these is selected by the filter no.1 (which has a low impedance at this frequency) and is fed back into the wheel at S_1 (an appropriate circuit in the aerial prevents reradiation at this frequency). This new frequency passing through the wheel now produces $(\omega_c - \omega_i) - \omega_i$ and $(\omega_c - \omega_i) + \omega_i$. The lower one at $\omega_c - 2\omega_i$ is again selected by the second filter and once more fed back into the wheel. A further pass through the wheel now produces $\omega_c - 3\omega_i$. If ω_i were chosen to be approximately $\omega_c/3$, this final frequency would be in the audio range. It is thus selected by the third filter and fed into the earpiece.

Goldschmidt experimented with another form of frequency convertor which was actually a modification of his high-frequency alternator. This was a rotary machine designed to produce carrier waves directly. Imagine that in such a device the stator coils were to be fed with an incoming high-frequency signal in such a way as to produce a rotating magnetic field. If the rotor were now spun at synchronous speed its coils would be keeping up with the field and no voltage would be induced in them. If, however, slip were allowed to occur so that the rotor went more slowly than the field a low frequency voltage would be produced in its coils, the exact frequency depending on the extent of the slip.[33] By way of example we may quote the figures given by the inventor himself, and quoted by Erskine-Murray;[34] with a slip of 1·67% and a carrier frequency of 30 000 Hz the frequency of the voltage induced in the rotor would be 500 Hz. This would naturally be audible in a telephone receiver connected to the rotor windings by means of slip-rings. Although the idea was basically sound, it ran into some difficulties due to remanent magnetism in the iron of the machine causing nasty noises in the telephone receiver.[35-36]

As previously stated, the tickers and tone-wheel were really designed to be used with systems which had a steady continuous carrier during the course of an individual Morse symbol. Many authors reported that it was possible also to use the method for reception of spark generated signals,[37] although there was some divergence of opinion and experience here. W.H.Marchant,[38] for instance, seemed unconvinced for he stated quite clearly that 'the method labours under the disadvantage that it is not able to receive signals from the ordinary spark transmitters which give out damped and discontinuous oscillations'. One wonders why he experienced difficulty in the matter. Could it possibly have been that, in the particular spark transmission he tried to receive, the individual bursts of oscillation were so highly damped that they performed relatively few oscillations before dying away? This could have meant that the ticker did not have sufficient time to build up resonance and produce an audible sound in the earphone. Eccles, too, wrote that with the tone-wheel 'damped waves (such as those emitted from a spark station) are not heard as a note, only as a noise'. An important factor may well have been the frequency stability of the transmitter. If the r.f. carrier were of truly constant frequency then a tone-wheel would have produced a good clear audible note. A spark transmission consisted of bursts of damped r.f. The burst frequency would have depended on the precise instant of sparking and the carrier frequency within the bursts may also have been of doubtful frequency stability. Even with the tone-wheel, and particularly with the ticker, this irregularity in the transmitted signal may have been sufficient to degrade the output into a noise as Eccles suggested. Such a noise would have been less easy to pick out in the presence of other interfering noises and transmissions.

It will have become apparent to readers who are familiar with today's radio techniques that we are now beginning to deal with heterodyne reception in its modern sense of 'mixing' the incoming signal with a locally produced carrier. The tickers and tone-wheel must be regarded as the crude mechanical forerunners of the true heterodyne principle, and it will be as well to say a little about the history of that means of reception before concluding this Chapter.

Credit for the idea is given to Fessenden, and it is he who coined the word 'heterodyne' to describe the process of reception by beats.[39-41] His first patent on the subject was granted in 1902 and in this he suggested transmitting Morse signals simultaneously on two carriers differing in frequency by a small amount. The arrangement to be used at the receiver is shown in Fig. 8.12*a*. The two incoming carriers were picked up, each one by its own tuned aerial system, and the aerial currents were made to flow through two coils wound on a common laminated core. Beating would occur between the two oscillatory magnetic fields with the result that an audio tone would be heard from the thin iron diaphragm situated near the end of the core. The attractive force on a diaphragm in this situation is independent of the direction of the field in the core so that the diaphragm was effectively acting as a full-wave rectifier.

This suggestion of Fessenden's seems to have lain dormant for some years and it was not until about 1907 that an active interest was shown in the subject once more.[42-46] It was soon realised that it was not necessary to transmit two signals and that one of them could be replaced by a locally generated oscillation at the receiver. One means of doing this

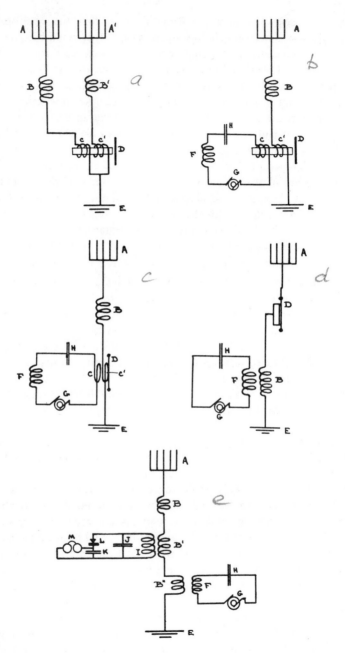

Fig. 8.12   *Various heterodyne receiver circuits proposed by Fessenden*
[Hogan, J.L.: *Proc. IRE*, **1**, 1913, p.87; British Patent 17 704, 1902]

was by the simple modification of Fig. 8.12*b*. Another way was to use a special telephone, as in Fig. 8.12*c*. It has one fixed coil to which is applied the locally generated carrier, and another coil attached to the diaphragm and connected in the aerial circuit. The beating effect between the two was made audible by the interaction of the fields of the two coils. Marius Latour[47] pointed out that a telephone of this sort was first used by Leblanc in 1886 in order to make audible an alternating current of ultrasonic frequency. The actual construction of Fessenden's telephone is shown in Fig. 8.13.

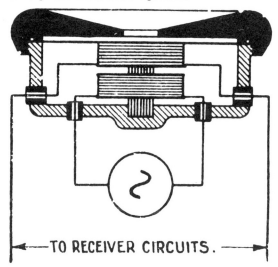

Fig. 8.13   *Fessenden's heterodyne receiver*
[Corresponds to the arrangement of Fig. 8.12*c*]
[Stanley, R.: *Textbook on wireless telegraphy* (Longmans, 1919), p.327]

Other possible circuit arrangements are shown in Figures 8.12*d* and *e*. In *d* the 'electrodynamometer' telephone is replaced by an electrostatic receiver, the thin diaphragm being connected to the aerial and the backplate to the locally generated carrier. In Fig. 8.12*d* we have a circuit of quite modern appearance where the beat frequency is extracted with the aid of a diode rectifier.

The apparatus of Fig. 8.14, also due to Fessenden,[48] is rather appealing. The beat frequency produced at the diaphragm (27) by interaction of the local and received currents in the coils (28 and 26) was fed straight into the horn of a recording cylinder phonograph. The whole idea was that the messages were to be sent at a very high rate with the aid of a punched-tape reader at the transmitter. This would be recorded on the wax cylinder and would afterwards be played back at a much slower rate enabling the operator, wearing the ear-tubes (42), to transcribe the information at his leisure.

Such then were the beginnings of the heterodyne system, one of the most important innovations in the whole field of radio, the full potential of which would only be realised with the development of convenient methods of generating stable continuous r.f. carrier waves.

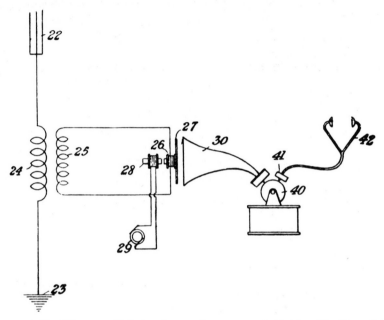

Fig. 8.14   *Phonographic heterodyne receiver arrangement proposed by Fessenden*
[British Patent 20 005, 1908]

References

1   GOLDSMITH, A.N.: British Patent 149 282, 1920
2   ARCO, G.G.: *Electrician,* **59**, 1907, p.566
3   TESLA, N.: British Patent 11 293, 1901
4   POULSEN, V.: British Patent 18 037, 1905
5   PEDERSEN, P.O.: British Patent 14 792, 1906
6   POULSEN, V.: British Patent 4 801, 1907
7   POULSEN, V.: British Patent 7799, 1907
8   CAMPBELL-SWINTON, A.A.: *Electrician,* **72**, 1914, p.687
9   STANLEY, R.: *Textbook on wireless telegraphy* (Lonmans, 1919), Vol.1, p.326
10   STONE, E.W.: *Elements of radio communication* (Van Nostrand, New York, 1926), p.312, section 531

11 ECCLES, W.H.: *Handbook of wireless telegraphy and telephony* (Benn, 2nd edn., 1918), p.318

12 BLAKE, G.G.: *History of wireless telegraphy and telephony* (Chapman and Hall, 1928), p.96

13 TISSOT, C.: *Manuel élémentaire de télégraphie sans fil* (Challamel, Paris, 1918), p.159

14 MARCHANT, W.H.: *Wireless telegraphy* (Whittaker, 1914), p.113

15 ADAM, M.: *Course de T.S.F.* (Editions Radio–Home, Paris, undated), p.132

16 ROUSSEL, J.: *Wireless for the amateur* (Constable, 1923), p.59

17 EDELMAN, P.E.: *Experimental wireless stations* (Published by Edelman at Minneapolis, 3rd edn., 1915/16), p.175

18 Reference 11, p.421

19 BUCHER, E.E.: *Practical wireless telegraphy* (Wireless Press, New York, 2nd edn., 1918), p.278

20 Reference 10

21 *Wireless World*, 7, 1919/20 p.8

22 Reference 12, p.98

23 Reference 15

24 Reference 16

25 Reference 19, p.276

26 DE FOREST, L.: *Proc. IRE*, **1**, 1913, p.37

27 Reference 10

28 POWELL, S.M.: *Electrical Review* (London), **74**, 1914, p.730

29 Reference 19, p.284

30 Reference 11, p.318

31 Reference 9, p.328

32 Reference 11, p.319

33 GOLDSCHMIDT, R.: British Patent 8387, 1911

34 ERSKINE-MURRAY, J.: *Handbook of wireless telegraphy* (Crosby, Lockwood, 4th edn., 1913), p.210

35 Reference 11, p.321

36 Reference 28

37 Reference 2

38 Reference 14, p.115

39 FESSENDEN, R.A.: British Patent 17 704, 1902

40 FESSENDEN, R.A.: US Patent 706 740, 1902

41 HOGAN, J.L.: *Proc. IRE*, **1**, 1913, p.75

42 FESSENDEN, R.A.: British Patent 6203, 1907

43 RUHMER, E.: *Wireless Telegraphy* (Crosby, Lockwood, 1908), p.201

44 *Electrician*, **71**, 1913, p.486

45 Reference 12, p,74

46 FLEMING, J.A.: *Principles of electric wave telegraphy and telephony* (Longmans, Green, 3rd edn., 1916), p.705

47 LATOUR, M.: *Radio Review*, **2**, 1921, p.15

48 FESSENDEN, R.A.: British Patent 20 005, 1908

Chapter 9

Miscellaneous detectors

Anyone who has ever tried to organise the contents of a filing cabinet will know that it is invariably necessary to have a section labelled 'Odds and ends' in order to accommodate those items which are unlike any others, or which are not sufficiently numerous to make it worth creating a special compartment to receive them. So it is that we must have the present Chapter.

Perhaps we should deal first with the very minor group of detectors which operated on the principle of electrostatic attraction. In 1899 B.Starkey[1-3] published the results of some wireless experiments in which the receiver consisted simply of a thin needle made out of silver paper balanced on an earthed point, the whole arrangement being placed very near the end of the receiving aerial, as in Fig. 9.1. The induced radio-frequency voltage in the aerial caused electrostatic attraction and the needle swung around, indicating the presence of the signal. A slightly modified set-up where the foil needle was suspended between two plates was described by Bouasse[4] in his book *Oscillations Electrique* in 1924.

The only other detector in this very minor group was the instrument shown in Fig. 9.2 which was patented by Fessenden[5] in 1910. Two fine wires (*a* and *b*) are held under tension in an evacuated tube and are connected one to the aerial and the other to earth at the receiver. Electrostatic attraction (which is independent of the polarity of the signal at any given time) causes them to move together slightly when a signal is present. This movement can be observed or recorded photographically by the use of an appropriate optical system. As a refinement,

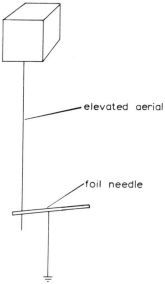

Fig. 9.1   *Starkey's electrostatic detector*

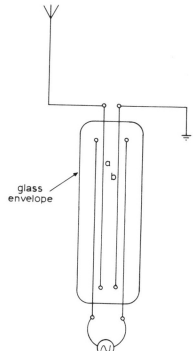

Fig. 9.2   *Fessenden's electrostatic receiver tube*

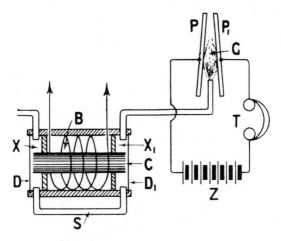

Fig. 9.3   *Horton's flame amplifier*
[Blake, G.G.: *History of radio telegraphy and telephony* (Chapman & Hall, 1928), p.209]

two other wires are stretched parallel to the first pair and a local oscillator is applied between them. The two electric fields then interact in such a way that beats are produced, and this produces beating which makes the photographic record more easily visible.

With the possible addition of the electrostatic telephone used by Fessenden in one of his heterodyne arrangements (Fig. 8.12*d*), this brings to an end the brief list of detectors which worked on the principle of electrostatic attraction.

In his presidential address to the London Wireless Society (later the RSGB) in 1914, A.A.Campbell-Swinton[6] gave an account of some interesting contrivances which made use of a sensitive flame to indicate the presence of a signal. A gas flame issuing from a very fine jet may be adjusted to a critical condition by controlling the gas pressure. It will then burn with a long steady flame, but if any disturbance in the pressure of the surrounding air should then occur the gas flow becomes turbulent and it will burn with a noisy ragged flame until quiet conditions prevail once more, when the long flame will return. The traditional way of demonstrating this effect in the school laboratory is to jangle a bunch of keys near the flame. In one of Campbell-Swinton's devices the gas was supplied to the jet through a 'manometric chamber'. This was simply a chamber with a thin paper diaphragm stretched across it included in the gas supply line. When a telephone connected to any one of the usual types of detector was held near it, it caused a minor fluctuation in the pressure of the gas supply which was sufficient to disturb the flame and to cause it to burn for a moment with a roaring sound. Morse signals were made audible in this way throughout a large lecture room.

At the beginning of Chapter 5 we reminded ourselves of the basic definition of the word 'detector', and pointed out that it was, for example, the coherer and earpiece *in combination* which actually performed the function of making the signal audible to a listener. Neither the coherer nor the earphone alone could achieve this. In considering Campbell-Swinton's gas flame device (and several others which follow) we may perhaps be pushing this definition a bit far since the detection had been carried out before the signal got to the flame. Such devices were relays rather than true detectors, but were useful and ingenious enough to merit inclusion here nonetheless.

In another of Campbell-Swinton's demonstrations the gas issued from a minute orifice in a glass jet above which was situated a sheet of metal gauze. The gas was ignited on the side of the gauze remote from the jet. The sensitive area was near the mouth of the jet, and a telephone receiver emitting clicks in this region would cause the flame to spread out across the surface of the gauze. If an opaque screen were placed around the flame the light could be made to appear and disappear as the flame burned high or spread out across the gauze. In yet another of Campbell-Swinton's highly original demonstrations the signal was made to open and close a valve in an air pipe. The air supply operated a whistle so that the dots and dashes were heard.

Another piece of apparatus which made use of a gas flame was due to Horton,[7] and this is illustrated in Fig. 9.3. The gas supplying the jet passes through the special chamber which contains a coil wound on a laminated magnetic core. Current from a crystal or some other detector flowing through the coil attracts the two thin iron diaphragms (D and D_1). The consequent changes in gas pressure cause the flame to rise and fall. Ionisation present in the flame allows current from the battery to pass between the two electrodes (P and P_1). As the flame moves, the conductivity varies and the signals are heard, magnified, in the headphones (T). This was referred to as 'Horton's flame amplifier'.

Other very sensitive indicators were the jet relays invented by Axel Orling.[8] These were primarily designed for use in cable telegraphy, but were also used in wireless reception. The basic principle is illustrated in Fig. 9.4. A thin stream of acidulated water falls from a fine glass tube. A light moving-coil is suspended nearby in a magnetic field, and a projecting spar fixed to this is arranged to be very near the jet. When current from the aerial rectified by a diode passes through the coil it rotates slightly and the spar is placed in the way of the stream of water and disturbs its smooth fall. In Orling's more refined models [9-10] a

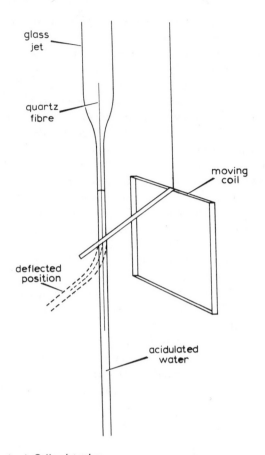

Fig. 9.4   Axel–*Orling jet relay*

fine quartz fibre passed through the centre of the jet; this was bent aside slightly by the movement of the spar and the jet of fluid was smoothly displaced without being broken up in any way, to land eventually on a different spot. The patent specified many different ways in which this movement could be utilised. In one version the jet normally fell on a conducting plate so that current could flow from a cell connected between the liquid in the tube and the plate. Deflection of the jet off the plate broke the circuit and the cessation of current was observed by means of a galvanometer. In the one shown in Fig. 9.5 the fluid itself was coloured ink which fell directly on to a moving strip of paper and the wavy line served as a permanent record of the received signal. In other arrangements the jet movement was made to disturb the balance of a Wheatstone bridge network.[11–12] For example, in its

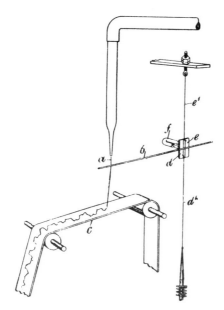

Fig. 9.5   *Orling's direct-writing ink jet relay*
[British Patent 18 001, 1911]

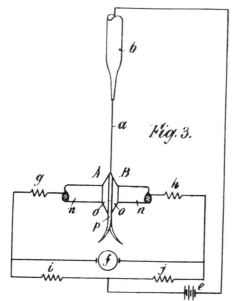

Fig. 9.6   *Jet relay with a bridge circuit to detect small movements of the fluid jet*
[Orling: British Patent 103 238, 1916]

undisturbed position the jet was made to fall symmetrically on to a wedge-shaped piece of glass sandwiched between two electrodes, as in Fig. 9.6. Equal quantities of fluid flowed down each side of the wedge. The two streams had equal resistances and these were incorporated as the ratio arms of the bridge which was then balanced. The slightest motion of the jet then caused unequal flow which resulted in severe imbalance in the bridge, indicated by the meter (*f*). A similar idea lay behind the system of Fig. 9.7. In this case the stream was divided between two porous 'horns' (A and B) incorporated in a bridge, and

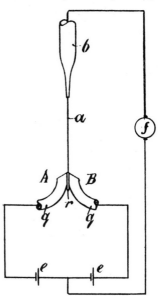

Fig. 9.7   *Jet relay with porous horns (q)*
[Orling: British Patent 103 238, 1916]

once again any small movement from the central position would produce imbalance.

Orling was not the only one to experiment with fluid jets. W.J.Lyons[13] produced the arrangement of Fig. 9.8. Normally the jet of acidulated water (15) flowed steadily and smoothly and struck the metal plate (16), thereby completing an electric circuit. When a pulse of current from a rectifier flowed in the coil (*j*) the tines of the tuning fork were set into vibration with the effect that the jet broke up into discrete particles and, of course, interrupted the circuit.

R.E.Hall[14] used an air jet in the manner shown in Fig. 9.9. Air under pressure normally flows out of nozzle A. In this stream there is a bolometer wire which carries a heating current from the battery E, and

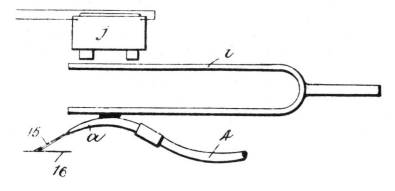

Fig. 9.8   *Lyons's jet relay*
[Lyons: British Patent 117 090, 1917]

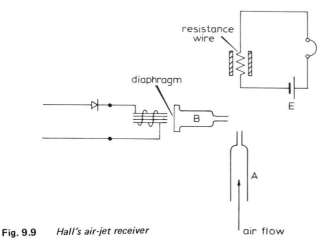

Fig. 9.9   *Hall's air-jet receiver*

it settles at some particular temperature. A telephone-type receiver has its diaphragm in connection with a chamber B whose only vent to the air is a small orifice which produces transverse air movement across the direction of the main jet. The telephone receiver is connected to a receiving circuit containing any one of the detectors which have been described, and the pulse of current causes the diaphragm to puff a jet of air across the jet, knocking it off course, as it were. There is a momentary change in the resistance of the bolometer wire which is made audible by headphones in the heater circuit.

Orling[15] also suggested the use of an air jet as in Fig. 9.10. The wire (3) forms part of a balanced bridge. The fine air jet is normally allowed to play on the point 3^0 which, being at the centre does not cause any

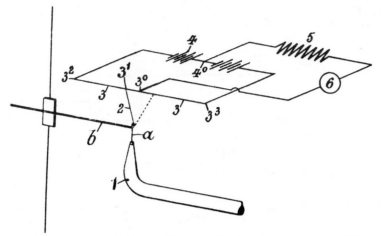

Fig. 9.10   *Orling's air-jet relay with bridge circuit to detect movement of jet*
[British Patent 18 001, 1911]

disturbance. If motion of a moving coil is made to deflect the jet to a point such as 3^1, one side of the bridge will be cooled relative to the other and the balance will be upset by the resulting change in resistance.

One of the most useful of all the relays was the mechanical model constructed by S.G.Brown[16-17] (Figs. 9.11 and 9.12). A horseshoe magnet lying horizontally had two vertical extensions to its pole pieces, to which were fixed two smaller extensions bearing the coils H. The coils K were wound around the larger pieces and were connected in series with a meter (D), battery (C) and headphones (T). Near the smaller polepieces there was a strip of springy Invar steel bearing a contact plate (O) which just touched a fixed point contact (M), both plate and point being made of highly polished osmium/indium alloy. A tiny drop of oil was placed between them. Current from a crystal detector flows in coils H and causes the contact pressure, and hence the resistance, to alter. A much larger variation of current then occurs in the headphone circuit. The whole thing was arranged rather cleverly to be self adjusting. Once the contact spacing was set the circuit reacted to changes in such a way as to tend to maintain this spacing. If, for some reason, the spacing were to increase slightly, the resistance would increase so that less current would flow from the battery around the coils K. The pull of the magnet on the spring strip would be decreased, tending to reduce the spacing to its former value. It was undesirable for the variations one was trying to detect to behave in this way, and so the windings K were surrounded by two shorted turns so that they had no inductive effects as far as a.c. variations were concerned, but the steady-state d.c. balancing would continue as previously described.

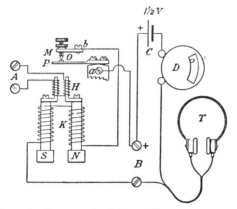

Fig. 9.11   *Brown's relay; a mechanical amplifier*
[*JIEE,* **45**, 1910, p.590]

This was a very sensitive device and, to quote Brown himself, 'In a wireless receiving station, messages, the very existence of which was not even suspected owing to their extreme feebleness when listened for under former conditions, with the relay in circuit were easily read.' In fact it was far more than a simple on/off relay, as it could respond

Fig. 9.12   *S.G.Brown's relay*
[*JIEE,* **45**, 1910 p.592]

gradually to gradual changes in the input signal and was thus usable with voice-frequency currents. It was nothing more or less than a mechanical amplifier, a very useful thing in the days before valve amplifiers became commonplace. There were several versions made. One which was particularly useful had the point contact replaced by a small box containing carbon granules, their compression and rarefaction providing a variable resistance as in today's telephone. It was used as a telephone repeater and even as a means of providing loudspeaker reproduction without valves in the early days of broadcasting.

In the opening paragraph of this book the remark was passed that the physicist has rather neglected the sense of taste. The reader may, however, be interested to learn that one Arthur A. Isbell suggested using this sense for the reception of radio signals. The idea was investigated at length by A.N. Goldsmith and E.T. Dickey,[18] their results being presented before the Institution of Radio Engineers, New York, in 1921. The general idea was that two small silver electrodes were either placed against the tongue, or else one was placed at the back of the upper lip and the other against the tip of the tongue. The audiofrequency signal obtained after detection was amplified (a valve amplifier was available by that date) and applied to the electrodes whereupon a sensation of taste was produced. Their researches showed that it was possible to read Morse signals in this way up to a speed of 5 – 10 words per minute, but there were several rather offputting disadvantages.

First of all the sensation produced was a 'sour stinging taste which was far from pleasant.' They also noted the quite interesting fact that 'a slight difference existed between the taste sensation produced by a 240-cycle spark transmitter and a 500-cycle spark transmitter.' Another serious disadvantage was that when the electrodes were placed at the back of the upper lip and the tongue there was a marked tendency to produce the sensation of toothache in the front teeth. Another curious phenomenon was that application of a pulse also caused a momentary contraction of the irises in the eyes; the sensation to the user was as though the lights in the room dimmed with each pulse. The overall conclusion was that although reception in this way was possible it was a vastly inferior method compared with the use of the senses of sight and hearing.

We shall now turn our attention to the work of Dr. Lefeuvre, Professor of Physiology at Rennes in Brittany.[19-20] His receiver, which was referred to in the literature as the 'Physiological Detector' is shown in diagrammatic form in Fig. 9.13. It will be observed that this was an adaptation of the well-known experiment attributed to Galvani in

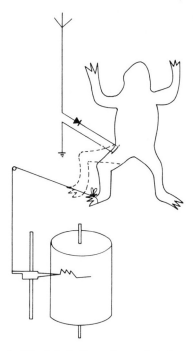

Fig. 9.13   *Lefeuvre's 'physiological' receiver using the well known electrical sensitivity of the frog's leg*

which the muscle of a frog's leg is made to twitch in response to an electrical stimulus. The frog was mounted vertically on a board and its sciatic nerve was carefully exposed. It was connected to the aerial circuit as shown. On the receipt of oscillation in the aerial circuit, the oscillatory current was rectified (by an electrolytic detector actually) and applied to the nerve, causing the leg to jerk convulsively. The leg was connected by way of string and pulleys to a pointer moving along the smoked surface of a drum, the drum being rotated so that the movement of the leg was recorded. With this apparatus Professor Lefeuvre was able to receive time signals from the Paris transmitter over 200 miles away. H.R.B. Hickman[21] who duplicated these experiments using a crystal rectifier in place of the electrolytic cell reported that he was able to receive this same signal at a place 30 miles north of London, being a distance of some 290 miles from the transmitter.

Of course the onset of rigor mortis was a bit of a problem and Hickman apologised for the small magnitude of the traces he published, explaining that 'the muscle-nerve preparation had been in use for three hours before the photographs were taken or the amplitudes would have been greater.' This drawback he alleviated by the grue-

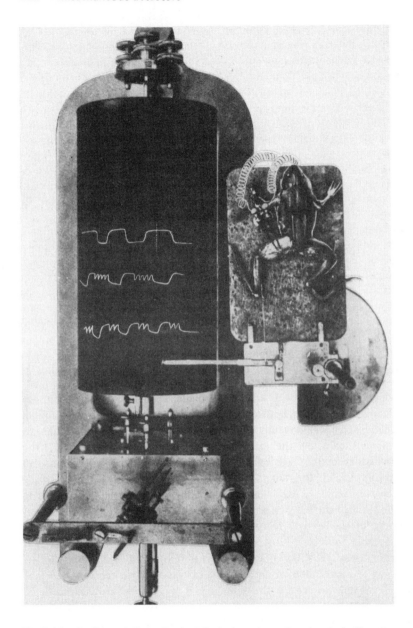

Fig. 9.14   *Professor Lefeuvre's physiological receiver using electrical effects in a frog's leg*
[Fleming, J.A.: *Principles of electric wave telegraphy and telephony*
[Longmans, 1916 p.547]

some expedient of 'employing the whole animal. The central nervous system (brain and spinal marrow) was destroyed, but as the action of the heart keeps up, the muscle will keep its senstiveness for a longer time than if it were detached from the body.' As Gradenwitz[22] explained in his account of these experiments 'frogs' legs are particularly well adapted for the purpose on account of their regularity of form, the long duration of their excitability and the special ease with which they (the nerves) can be separated from the body.' In case the reader should be tempted to wonder whether these various accounts were written on the first day of April, Fig. 9.14 shows a photograph of Lefeuvre's apparatus, and Fig. 9.15 the actual traces of the signal received from the Paris transmitter.

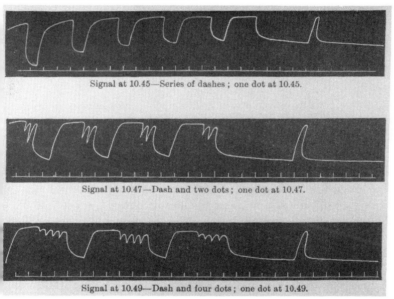

Signal at 10.45—Series of dashes ; one dot at 10.45.

Signal at 10.47—Dash and two dots ; one dot at 10.47.

Signal at 10.49—Dash and four dots ; one dot at 10.49.

Fig. 9.15   *Signals received from Paris with Lefeuvre's receiver*
[Fleming, J.A.: *Principles of electric wave telegraphy and telephony* (Longmans, 1916)]

One's reaction at this point is to be thankful that no one has ever gone so far as to conduct experiments of this sort with a human subject. However, one would be wrong in such a supposition, although one must hasten to reassure the squeamish reader that the human subject concerned was dead (just). In 1902 A. Frederick Collins published a paper entitled 'The effect of electric waves on the human brain,'[23] the aim of which was 'to verify, if possible, the casual observations long since made that approaching electrical storms manifested their presence in persons afflicted with certain forms of nervousness and other pathological conditions, though the storm influencing them might be many miles beyond, or even below the horizon.' Collins constructed a normal type of spark transmitter with Hertzian rod resonators and the fairly conventional form of receiver shown in Fig. 9.16. The two parts labelled B were needles armoured with sheaths of glass, and D was a telephone earpiece.

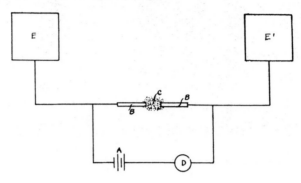

Fig. 9.16   *Receiving apparatus used by Collins during his investigations into the behaviour of the human brain as a coherer*
[*Electrical World and Engineer,* **39**, 1902, p.335]

Collins' first step was to obtain from his butcher the brain of a freshly killed mammal (type unspecified) into which he inserted the two needles in an attempt to find out whether it would act like a coherer. The brain is represented by the dotted mass C in Fig. 9.16. He hoped that the resistance between the probes would fall when oscillations were received, but his initial results were disappointing.

The next step was to use the brain of a cat, first of all a dead one, then in his own words, 'another cat under the influence of an anaesthetic, willingly lent itself (sic) to the subject for the investigations to be made on brain matter in the living state.' Excellent results were obtained, and he noted a vibrating movement at the base of the brain, 'a spasmodic twitching of the muscles just as a "shocking coil" would produce.'

Encouraged by these results he went on to lament 'let us turn to the human brain. A brain, when death has eliminated its spiritual affinity from the grosser physical form is of little worth to anyone except the anatomist. Yet clay though this wonderful organ is in death, it is a most difficult object to obtain.' He was in luck, for 'on a certain afternoon a magnificent specimen of a human brain came into my possession immediately after its removal from the cranium, and within hours of the death of the physical body.' Cutting a long

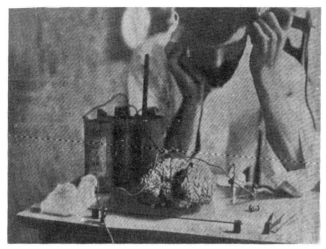

Fig. 9.17 *A.F.Collins listening to the 'cohesion of the human brain under the action of electric waves'* [*Electrical World and Engineer,* **39,** 1902]

story short, he obtained good results again, finding that the rust-coloured granular layer in the cerebellum was the most efficacious. The end of the tale reads like a thriller with a final twist to its plot, too good to be true:

> In conclusion, I intended to once more measure the resistence of the brain as a whole to ascertain if possible the relative differences in resistivity the deterioration 24 hours would produce. The needles were inserted in the opposite hemispheres of the brain and connected with the terminals of a Queen testing set. It was with some astonishment I found I could not balance the bridge arms. After an approximate balance, the needle suddenly swung to + without apparent cause, showing hundreds of ohms too much, and an immediate second deflection to the – showing hundreds of ohms too little, without the removal or insertion of a plug. This state continued for a few minutes, when a peal of thunder awakened me to the actual cause. A storm was approaching. As the storm approached, the deflections grew more and more pronounced, the needle quivering at either end of the scale alternately as though endowed with life. The very phenomenon I sought to verify with a 2 cm spark coil was here produced by the lightning itself . . . In these tests I was favoured with circumstances which, with me, might never occur again, for the reason that a fresh human brain was necessary, and that an electrical storm should be in progress when all was in readiness was quite remarkable.

Readers of delicate sensibilities are advised to refrain from looking at Fig. 9.17, which shows Mr. Collins conducting his grisly experiments.

The Editor of *The Electrician*, writing about Lefeuvre's frog's leg experiments[24] was either unaware of Collins's findings or incredulous, for he said 'perhaps those who write "scientific" articles for our daily contemporaries will see in this an explanation of the twitching which some folk feel at the approach of a thunderstorm. But it occurs to us that oscillatory current cannot in fact affect nerves and muscles, for if it could, then in spite of the 'skin effect' the neighbourhood of a large wireless telegraph station would be full of votaries of St. Vitus during the despatch of a message'!

References

1   STARKEY, W.B.: *Electrical Review*, 45, 1899, p.1026
2   STARKEY, W.B.: *Electrical World and Engineer*, 35, 1900, p.73
3   TURPAIN, A.: *Les applications pratiques des ondes éléctriques* (Carré et Naud, Paris, 1902), p.366
4   BOUASSE, H.: *Oscillations éléctrique* (Delagrave, Paris, 1924), p.280
5   FESSENDEN, R.A.: British Patent 11 155, 1910
6   *Electrician*, 72, 1914, p.687
7   BLAKE, G.G.: *History of radio telegraphy and telephony* (Chapman and Hall), 1928), p.209
8   ORLING, A.: British Patent 18 001, 1911
9   Reference 7, p.116
10 FLEMING, J.A.: *Principles of electric wave telegraphy and telephony* (Longmans, Green, 3rd edn., 1916), p.715
11 ORLING, G.: British Patent 103 238, 1916
12 ROUSSEL, J.: *Wireless for the amateur* (Constable, 1923), p.124
13 LYONS, W.J.: British Patent 117 090, 1917
14 HALL, R.E.: British Patent 144 250, 1920
15 Reference 8
16 BROWN, S.G.: *JIEE*, 45, 1910, p.590
17 Reference 10, p.717
18 GOLDSMITH, A.N., and DICKEY, E.T.: *Proc. IRE*, 9, 1921, p.206
19 *Electrician*, 71, 1913, p.93
20 Reference 10, p.540
21 HICKMAN, H.R.B.: *Electrician*, 71, 1913, p.143
22 GRADENWITZ, A.: *Electrical Review* (London), 71, 1912, p.820
23 COLLINS, A.F.: *Electrical World and Engineer*, 39, 1902, p.335
24 *Electrician*, 71, 1913, p.81

Chapter 10

And so to the modern era

The one great principle which gradually emerged from all these receiving methods was that of demodulation by rectification. It has already been explained in Chaper 7 how the thermoelectric detector merged almost imperceptibly into the rectifier, and the reader may recall that some of the electrolytic detectors of Chapter 4 were also found to possess the property of rectification. When Professor Lefeuvre required a unidirectional current to stimulate his frog's leg, it was to the electrolytic cell that he turned.

To many people the coherer too seemed to develop logically into the rectifier. A typical rectifier has a voltage/current characteristic of the general shape shown in Fig. 10.1. When the characteristic curve for a coherer was measured, it had a very similar general shape save that it was symmetrical as indicated by the dotted line. With both diode and coherer it was advantageous to apply a small bias voltage to take the device up to the bend in the curve, thereby improving its sensitivity to the required signal. The great difference between them was that the coherer needed to be tapped to restore it to the low-current state; but there again, there were some 'self-restoring' coherers which needed no such mechanical disturbance. A coherer in a receiving circuit allowed only a small current to pass from the local battery until the oscillations were applied, at which point it started to conduct. A diode connected in a similar circuit would conduct very little current while it was below the bend in the curve, but oscillations which pushed the operating point further up the curve caused proportionately larger currents to flow.

The similarity between the two devices was really very strong, and in fact W.H.Eccles in his book *Wireless telegraphy and telephony* fails to make a clear distinction between the two effects, regarding both as different manifestations of the same thermoelectric phenomena.[1] As

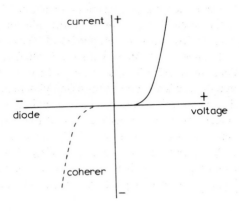

Fig. 10.1   *Similarity between the voltage/current characteristics of the coherer and the diode*

another example of the way that experimenters of the time regarded them as closely related devices; as late as 1919 the *United States Manual of Radio Telegraphy and Telephony* felt it necessary to point out that the crystal and the valve were self restoring.[2] Many other authors made the same point.[3]

It was the principle of rectification as embodied in the crystal and the thermionic diode which allowed the easy demodulation of an amplitude-modulated carrier and this, together with the possibility of amplification arising from the introduction of the 'grid' into the diode, paved the way for radio as we know it today. Since this book is concerned with early detection methods, this is really the point at which we shall leave the story, but it would be wrong to do so without saying just a little about these important developments. As may well be imagined, the literature on the subject of crystal and valve detectors is quite enormous and to do it justice would require a great deal of space, especially if all the various derivatives of the triode valve were to be considered. For this reason the essential points alone will be noted and a few references will be given to start off the interested reader who may wish to enquire more deeply.

The story of the crystal rectifier started in the 1870's with the discovery by Braun of the asymmetrical conduction properties of certain metal sulphides.[4] Other investigations and patents concerning the junctions of metals and crystalline substances followed in the early years of the 20th century[5-6] but many of these, although used as rectifiers, were regarded by their inventors as being thermoelectric devices as already mentioned. In the years between about 1906 and the 1914-18 War, great interest was shown in the phenomenon, and various investigators looked into the properties of every conceivable combination of substances.[7-19] Many pairs were found to behave as rectifiers, some being more effective than others, of course. Among the most popular materials used were molybdenite (molybdenum sulphide), galena (lead sulphide) and carborundum (silicon carbide). Many devices were marketed under trade names such as the Perikon detector [a chalcopyrites (mixed copper and iron sulphide)/zincite (zinc oxide)] combination, the Pyron [iron pyrites (iron sulphide)/silicon] and the Bronc Cell [tellurium/graphite]. It is interesting to note in passing that as early as 1907 the emission of light from a rectifying junction had been observed.[20]

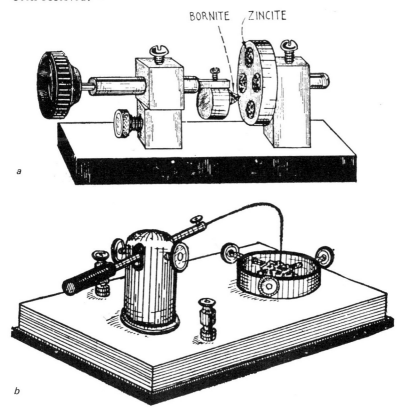

Fig. 10.2   *Typical adjustable crystal holders*
[Bucher, E.E.: *Practical wireless telegraphy* (Wireless Press, New York, 1917)

In some of these combinations the rectifying action took place between two pieces of crystalline material and holders such as that of Fig. 10.2*a* were designed to hold them and to achieve the best contact pressure. In those using the crystal/metal junction it was often necessary to find a sensitive spot on the surface of the crystal by searching around with a fine metal point. This was the famous 'cat's whisker'. The crystal mountings were many and varied, and those shown in Figs. 10.2*b* and 10.3 are simply typical examples.[21-23] Fig. 10.4 shows a type of holder designed to be plugged into a socket as a replacement for a diode valve. Fig. 10.5 shows a complete crystal receiver in which the cat's whisker and a spare are clearly visible. Many a schoolboy has built himself a crystal receiver and has spent pleasant, often frustrating, hours 'tickling' the crystal to find that elusive spot. Today's children are usually denied this pleasure since nowadays crystal sets sold in kit form contain a modern semiconductor device as the rectifying element,

Fig. 10.3 *Two crystal detectors*
(*a*)   With delicate screw-operated pressure adjustment
(*b*)   With several crystals on a rotating turret
[Science Museum photographs]

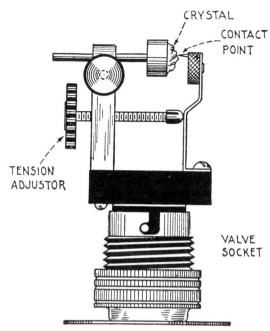

Fig. 10.4   *Crystal holder designed as a plug-in replacement for a Fleming diode valve*
[Bucher, E.E.: *Practical wireless telegraphy* (Wireless Press, New York, 1917), p.143]

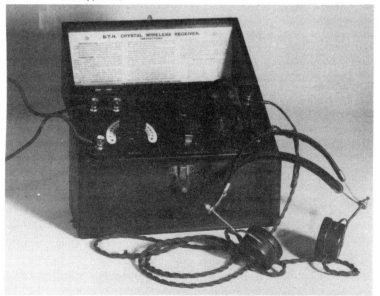

Fig. 10.5   *BTH crystal receiver with 'cat's whisker' and spare*

although crystals for cat's whiskers were sold until well after the Second World War.

The other line of development in rectification was the thermionic diode. Thomas A. Edison had noticed in 1883 that when a metal plate was inserted into the evacuated bulb of a filament lamp a current could flow between the filament and the plate across the apparently empty space. This phenomenon, often called the Edison effect, lay unused until 1904 when J.A.Fleming filed his patent for what he called the 'oscillation valve'.[24-27] This is rather a confusing name to the present-day engineer, who would assume that it was a valve capable of producing oscillations; it was actually a simple vacuum diode acting towards oscillatory current in the same way as a nonreturn valve behaves to water in a pipe (Fig. 10.6). This was not the only line of development as there were some experimenters who advocated the use of ions rather

Fig. 10.6   *Fleming's diode valve or 'oscillation valve'*
[Science Museum photograph]

than electrons *in vacuo*. As we have seen in previous Chapters of this book, ions exist in a gas flame, and Lee De Forest[28-29] devised the system of Fig. 10.7 to act as a rectifier. It probably worked reasonably well since Eccles referred to it as 'having properties of a most felicitous and commodious character'.[30] G.Leithäuser[31-32] in 1912 actually proposed using a gas flame seeded with potassium salts as the conducting medium in order to overcome the problem often encountered with the 'hard' evacuated diode, namely the deterioration with time of the vacuum.

By a series of patents Lee De Forest transformed his Bunsen-burner arrangement into a device which was virtually the same as Fleming's diode, and to which he gave the name 'audion'.[33-37] Long legal wrangles then insued, the eventual outcome of which was a decision in the courts in favour of Fleming.[38] Readers who require further details of this protracted affair can do no better than to consult G.F.Tyne's detailed book *The saga of the vacuum tube.*[39] Fleming's version of events may be read in his book *The thermionic valve and its developments in radio telegraphy and telephony.*[40]

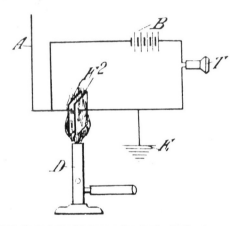

Fig. 10.7   *De Forest diode-type detector using the ionisation in a gas flame*
[British Patent 5 258, 1906]

In 1906–7, De Forest made a great step forward by introducing a third electrode, the grid, in between the filament and the anode of the vacuum diode. Fleming himself acknowledged De Forest's right to be considered as the inventor of this 'triode' valve which would not only rectify the signal but also amplify it — an advance of far-reaching significance. In the United States, the name 'audion' was often applied to this three-electrode valve, although the word was not used very much on this side of the Atlantic. At a later stage, the General Electric Company of America tried to rechristen it the 'pliotron' on the grounds that its vacuum was more perfect than that employed in De Forest's and Fleming's valves. A letter from De Forest[41] scathingly refers to 'the Graeco-Schenectady name "pliotron", an appellation with which the General Electric Co. has for some time seen fit to disguise its use of the "audion"'. (Schenectady was, and is, the home of the General Electric Co.). Other names were used by various people from time to time; these included the 'kenotron', 'dynotron', 'oscillion' and 'thermotron'.[42]

Just as the crystal rectifier had ousted all the other detectors because of its reliability and convenience, so the even better performance of the diode enabled it to replace the crystal in its turn. In the 1911 edition of the *Manual of wireless telegraphy for the use of naval electricians*[43] S.S.Robison wrote, 'There are but two types of detector now in general use, crystal or rectifying detectors, and the electrolytic. Coherers and microphones are practically obsolete and comparatively few of the audion or valve detectors have been installed'. In the 1913[44] edition he changed this to 'Only one type of detector is now in use, the crystal. The electrolytic is used as a standard of comparison. Coherers and microphones are practically obsolete and comparatively few magnetic and audion or valve detectors have been installed'. By 1919[45] the wording had become simply 'The detectors now in general use are the crystal or rectifying detector and the audion' – the other types had obviously faded away by this date.

Our own *British Admiralty handbook of wireless telegraphy* for 1920 states,[46] 'Crystal detectors are being replaced by valve detectors which are more stable, easier to adjust and generally more satisfactory'. By the 1925 edition it was clear that the valve had won, 'replacing the crystal for all ordinary purposes'.[47] Who would have guessed that thirty years or so later the semiconducting crystal and its derivatives would take their revenge in fullest measure!

The present author has lectured on the subject of early Hertzian wave detectors on numerous occasions, and during question time at the end he is invariably asked the question 'how sensitive were these detectors?'. This is a very difficult question to answer as information on the matter is very hard to come by. The literature abounds with statements such as 'the such-and-such detector was found to be as sensitive as a good filings coherer' or 'it was found to be reliable in operation and possessed good sensitivity'. Some authors tried to be a little more objective in their assessments and make statements such as 'using the so-and-so detector, signals were easily obtained at a distance of x miles from the transmitter'. Usually no information is offered as to the power of the transmitter, and even when it is, the specification states the length of the transmitting spark, a measurement impossible to translate into modern terminology. Details of the receiving aerial are seldom given, and it is not possible to assign any quantitative value to these various statements.

Every attempt has been made wherever possible throughout the course of this book to give some general indication of the performance and effectiveness of the devices described. It is only in the case of the thermal detectors that really meaningful figures were available. Two authors did attempt a comparative assessment of the performance of various classes of detectors; their data are reproduced in Tables 10.1 and 10.2.

Edelman[48] uses as his standard for comparison the energy (in ergs; 10^7 ergs = 1 joule) required to make the detector indicate reliably the arrival of a Morse dot. The rectifying detector based on silicon comes out best, although the electrolytic detector is also seen to be very sensitive. Fessenden[49] is reported to have said that coherers needed 1 to 4 ergs to produce an indication.

Table 10.1   *Edelman's comparison of detectors (1915/16)*
[EDELMAN, P.: *Experimental wireless stations* (published by Edelman at Minneapolis, 1915/16) p.160]

TABLE OF DETECTORS–SENSITIVENESS

Type of detector	Energy required to operate in ergs. per dot.
Electrolytic	0.003640–0.000400*
	0.007 §
Silicon	0.000430–0.000450*
Magnetic hysteresis detector	0.01 §
Hot-wire barretter	0.08 §
Carborundum	0.009000–0.014000*

    *   According to Pickard.
    §   According to Fessenden.

S.M.Powell's analysis[50] shown in Table 10.2 is rather more comprehensive, although five years earlier, and tries to assess the relative importance of such factors as automatic decoherence, speed of operation etc. by assigning marks according to a weighted scale. The reference to selenium, by the way, is purely speculative. He says in his explanatory text 'the phenomena of light sensitivity of selenium are well known and strongly pronounced and there seems no doubt that selenium is affected to a certain extent by electromagnetic and electrostatic waves, as well as by certain other invisible radiations. There is thus a possibility of a successful selenium detector being evolved though such is not yet available.' In other words, since selenium is sensitive to light he saw no reason at all why it should not also be sensitive to radio waves. One wonders how he could possibly have arrived at the scores for such a

Table 10.2

Comparing Various Types of 'Detectors'

Type of detector	Properties												Suitable for what purposes	Remarks, authorities, etc.
	Adjustability		Sensitivity				External effects				Quantitative in action dependent on maximum p.d.			
	Automatic decoherence	Tuning	Quick acting	To weak signals	To strong signals	Resistance	Vibration	Field	Ease of Construction and erection	Ease of pro-curing material	Total			
Maxima	5	5	5	5	3	4	4	4	2	1	43			
Coherer	0	0	5	4	3	2	5*	4	1	1	28	p.d.	Medium distance wireless telegraphy	
Magnetic	5	3	4½	5	2	3	4	3½	1	1	36	p.d.	Long-distance telegraphy; shorter distance telephony	Personal experience
Electrolytic	5	3	5	5	3	2	4	4	2	1	39	Ct.	Long or short-distance telegraphy and telephony	
Thermo-electric	5	5	4½	4½	3	3½	3	4	2	½	39	Ct.	Ditto, but less satisfactory for telephony	
Ordinary Rectifier	5	4	5	2	2	2½	4	4	2	½	36	p.d.	Suitable for experimental telegraphy	
Walter's Ta-Hg ditto	5	5	4	2	3	2½	4	4	2	1	37½	p.d.	Suitable for short-distance work or for exceptionally strong long-distance work	Walter, Eccles, Fessenden and others
Selenium	3	2	2	4	3	2	3	4	0	½	27½	?	Not likely to be of great use	No successful selenium detector yet known. Figures tentative
Vacuum 'valves'	5	3	4½	4	2	2½	4	3½	0	1	31½	p.d.	Suitable for general work, but no advantage in using in preference to electrolytic or thermal	Fleming's results hardly confirmed by other investigators

**Vibration is a very marked advantage in the case of coherers; hence the overmarking.

Powell's comparison of detectors [1911]
[*Electrical Review*, **68**, 1911, 13th Jan., p.74]

nonexistent detector! A point of interest which emerges from his comments is that attempts were being made at that stage to use existing detectors for telephony using an amplitude-modulated carrier. Some of them were able to give just a little more than a yes/no indication, and to a very limited extent could respond in varying degrees depending on the strength of the signal. In the years between 1904 and 1910, accounts of wireless *telephony* experiments reported in the magazine *The Electrician*[51-56] tend to concentrate on the production of the modulated carrier and are frustratingly vague about detection methods employed at the receiver. It is clear that thermal detectors were used, although there are hints that magnetic and electrolytic methods were also tried. From about 1908 onwards, experimenters seem to have relied increasingly upon rectification methods using crystals and thermionic tubes.[57-58]

The other notable feature of Powell's table is that in 1911 the conclusion was still that the vacuum diode was of 'no advantage in preference to electrolytic or thermal'. In the text of his paper he quotes Fleming (May, 1909) as having given the order of sensibility (sensitivity) as oscillation valve, magnetic detector, electrolytic and carborundum, but Powell obviously disagrees and says that this was 'by no means exhaustive or even universally true, so far as it goes'. The diode valve had clearly not yet made its great impact.

As has been remarked in a previous section, sensitivity was not everything; too sensitive a detector would tend to respond in an unwanted way to atmospheric disturbances, impairing reception of the wanted signal. In fact, the concepts of detection threshold and signal-to-noise ratio were beginning to dawn in a very practical way. In 1905 Walter[59] wrote that 'Captain Ferrié has several times reiterated the view that great sensitivity is not so desirable, and he advocates the use of greater power at the transmitter with a less sensitive detector'. This was the voice of practical experience.

This, then has been the story of the development of the essential part of the wireless receiver, that section which converts the high-frequency signal into a form which our senses can perceive. It has been an account of the ingenuity and inventiveness which were brought to bear on the problems encountered in the birth and development of radio, that marvellous invention which nowadays we take so much for granted; which amuses, instructs and informs us and which is such a vital link in our complex transportation systems, helping to ensure that wherever we may be, on land, sea or in the air, we shall have a speedy relief to our necessities if we are in danger, and a safe return to our native land should we so desire it.

References

1   ECCLES, W.H.: *Wireless telegraphy and telephony* (Benn, 2nd edn., 1918), p.270ff

2   ROBISON, S.S. *et al.: Manual of radio telegraphy and telephony* (US Navy Publication, 5th edn., 1919), section 189

3   BLAKE, G.G.: *Model Engineer and Electrician*, 19, 1908, p.515

4   BRAUN, F.: *Pogg. Ann.*, 153, 1874, p.556

5   BELLATI, M., and LUSSANA, S.: *J. Soc. Telegraph Eng.*, 18, 1889, p.202

6   BRANDES, H.: *Electrotech Zeitschrift*, 27, 1906, p.1015

7   ECCLES, W.H.: *Proc. Phys. Soc.* (London), 25, 1913, p.273

8   *Wireless World*, 1, 1913/14, p.240

9   TUTTON, A.E.H.: *Wireless World*, 1, 1913/14, p.232

10 WORRALL, H.T.: *Wireless World*, 2, 1914/15, p.434

11 PICKARD, G.W.: British Patent 2943, 1908

12 ROUSSEL, J.: *Wireless for the amateur* (Constable, 1923) p.32ff

13 DOWSETT, H.M.: *The Radio Review*, 2, 1921, pp.582 and 642

14 *Electrician*, 68, 1912, p.953

15 *Electrician*, 63, 1909, p.736

16 COURSEY, P.R.: *Proc. Phys. Soc.* (London), 26, 1914, p.97

17 PIERCE, G.W.: *Phys. Rev.*, 25, 1907, p.31

18 TORIKATA, W.: *Electrician*, 65, 1907, p.940

19 WALTER, L.H.: *Electrical Engineering*, 4, 1908, p.247

20 ROUND, H.J.: *Electrical World*, 49, 1907, p.308

21 BUCHER, E.E.: *Practical wireless telegraphy* (Wireless Press, New York, 2nd edn., 1918), p.139ff

22 Reference 1, pp.280 and 281

23 EDELMAN, P.E.: *Experimental wireless stations* (published by Edelman at Minneapolis, 3rd edn., 1915/16) p.161ff

24 FLEMING, J.A.: British Patent 24 850, 1904

25 *Electrical Magazine*, 6, 1906, p.460

26 FLEMING, J.A.: *Proc. Roy. Soc.* (London), 74, 1905, p.476

27 FLEMING, J.A.: British Patent 13 518, 1908

28 DE FOREST, L.: British Patent 3 380, 1907

29 DE FOREST, L.: British Patent 5258, 1906

30 ECCLES, W.H.: *Electrician*, 60, 1908, p.588

31 LEITHÄUSER, G.: *Physik Zeitschrift*, 13, 1912, p.892

32 *Electrical Review* (London), 72, 1913, p.450

33 DE FOREST, L.: British Patent 1427, 1908

34 DE FOREST, L.: *Electrician*, 72, 1913, pp.285, 274, 377, 659 and 660

35 ARMSTRONG, E.H.: *Proc. IRE*, 3, 1915, p.215

36 DE FOREST, L.: *Electrician*, 78, 1916, pp.505, 327 and 477

37 WALTER, L.H.: *Proc. Roy. Soc.* (London), 81A, 1908, p.1

38 *Wireless World*, 4, 1916, p.643

39 TYNE, G.F.J.: *Saga of the vacuum tube* (H.W.Sams, 1977)

40 FLEMING, J.A.: *The thermionic valve and its developments in radiotelegraphy and telephony*, (Wireless Press, 1919), Chap. 3

41 Reference 36

42 STONE, E.W.: *Elements of radio communication* (Van Nostrand, 1926), p.331

43 ROBISON, S.S.: *Manual of wireless telegraphy* (US Navy, 2nd edn., 1911), p.128
44 ROBISON, S.S.: *Manual of wireless telegraphy* (US Navy, 3rd edn., 1913), p.132
45 Reference 2
46 *Admiralty handbook of wireless telegraphy* (HMSO, 1920), p.264
47 *Admiralty handbook of wireless telegraphy* (HMSO, 1925), p.286
48 Reference 23, p.160
49 *Electrical Magazine*, 2, 1904, p.166
50 POWELL, S.M.: *Electrical Review (London)*, 68, 1911, p.74,11
51 *Electrician*, 65, 1910, p.1066
52 FESSENDEN, R.A.: *Electrician*, 59, 1907, p.9
53 FESSENDEN, R.A.: *Electrician*, 58, 1907, pp.675 and 710
54 FESSENDEN, R.A.: *Electrician*, 61, 1908, pp.786 and 867
55 *Electrician*, 53, 1904, p.991
56 CARLETTI, A.: *Electrician*, 62, 1909, p.609
57 *Electrician*, 60, 1908, p.978
58 *Electrician*, 63, 1909, p.736
59 WALTER, L.H.: *Electrical Magazine*, 3, 1905, p.513

Select bibliography

Readers who require further general information on the subject of detectors are referred to the following books:

FLEMING' J.A.: 'The principles of electric wave telegraphy and telephony' (Longmans, Green and Co, London. Various editions from 1906 onwards)
BLAKE, G.G.: 'History of radio telegraphy and telephony' (Chapman and Hall, London, 1928. Also available in modern reprint, Arno Press, New York, 1974)
ERSKINE-MURRAY, J.: 'A handbook of wireless telegraphy' (Crosby Lockwood, London, various editions from 1907 onwards)
MAZZOTTO, D.: *Wireless telegraphy and telephony* (Whittaker, 1906)
SEWALL, C.H.: *Wireless telegraphy* (Crosby-Lockwood/van Nostrand, 1904)
BOULANGER, J. and FERRIÉ, G.: *La telegraphie sans fil et les ondes electriques* (Berger-Levrault, 1907)
MONIER, E.: *La télégraphie sans fil* (Dunod et Pinat, various editions)
TURPAIN, A.: *Les applications des ondes électriques* (Carre et Naud, 1902)
The journal *The Electrician* is also a most valuable source of information.
The various other texts mentioned in the References are also full of interesting information, but the reader may find them a little more difficult to obtain.

Index

Numbers with decimal points refer to
diagrams

thermal effects on coherers, 59
thermal effects in electrolytes, 74,
  154
thermal explanation of rectification,
  169
thermocouple detectors, 164ff, 169
thermogalvanometer, 159, 7.12
thermophone, Eccles, 162, 7.14
thermotron, 212
thin-film detectors, 123ff
Thomson, E., 1, 86
Thomson, J.J., 5
Thomson, W., 4
tickers, 173ff
Tieri, L., 119, 5.34, 5.35
Tissot, C., 34, 47, 98, 101, 153,
  3.40, 5.17
Tommasina, T., 45, 55, 68, 3.37,
  3.49
tone wheel, 179, 8.10, 8.11
tongue, use for reception, 198
toothed wheel, 49, 172, 177, 8.8, 8.9
torsional-magnetic detectors, 119
transatlantic transmission, 56, 5.1
transmitters, 7, 1.5
trigger-tube, 12, 2.4
triode, 211
tripod coherer, 21, 3.3, 3.4, 3.42,
  3.43
tuning, 11, 20, 2.3, 3.1
Turpain, A., 15, 37, 2.9

Valve, thermionic, 8, 210ff
Vanni, G., 138
Varley, S.A., 18
vaseline/mercury mixture, 126
vibration,
  effect on coherer, 29
  effect on electrolyte, 81
  with oil film, 128
Von Lang, V., 18

Walter, L.H.,
  hysteresis detector, 113, 5.30
  on torsional detector, 121
  tantalum detector, 57, 3.51
  thin film detector, 132, 6.8
water jets, 134, 6.9, 6.10
wave responder, 6
wax, in coherers, 59

welding in coherers, 60
Wheatstone bridge, 151, 153, 192,
  195
wheel coherer, 128, 148, 6.3, 6.4
Wiedemann effect, 121
Wilson, E.,
  hysteresis detector, 98, 111, 5.27
  oil film detector, 126, 6.1
  remagnetising circuit, 93, 5.8, 5.9
Wollaston wire, 70, 151, 153

Zakrzewski, C., 80, 4.12
Zehnder, L., 12, 2.4